THE SCIENCE OF DESIGN

GORDON L. GLEGG

Consulting Engineer and Lecturer in Engineering
University of Cambridge

CAMBRIDGE
AT THE UNIVERSITY PRESS
1973

Published by the Syndics of the Cambridge University Press
Bentley House, 200 Euston Road, London NW1 2DB
American Branch: 32 East 57th Street, New York, N.Y.10022

Library of Congress Catalogue Card Number: 73–80471

ISBN: 0 521 20327 9

Printed in Great Britain
at the University Printing House, Cambridge
(Brooke Crutchley, University Printer)

CONTENTS

PREFACE

Longwindedly, but more accurately, this book could have been entitled 'The General Principles of Scientific Research for Obtaining Data for Engineering Design'.

In the contemporary world of engineering, the designer tends to become a specialist while needing to become a general practitioner. Knowing more and more about less and less, he should be learning more and more about more and more.

This is especially true when he finds that to complete a design he lacks some essential data that can only be found by research.

Now being a good designer does not automatically mean that he is gifted at doing research. Nor does it automatically mean that he is not gifted at it. Neither he, nor the Company or University that employs him, know the answer, and much water may flow under the bridge and much money flow out of the bank before they both can be sure what his aptitude is.

The intention of this book is to ensure that whatever the state of the reader is when he begins reading it, he will, at least, be slightly suitable for carrying out research responsibility by the time he reaches the end of it.

For the old hands I hope a realistically illustrated philosophy based on over forty years experience of research may prove a pattern that will add significance and order to their own.

Finally, this book is one in a series of three, which touch without overlapping, and are intended to cover the whole spectrum of engineering design. Although they can be read independently and in any order, the logical sequence is first that of the present book, dealing with research; next is *The Selection of Design* which covers inventive techniques. Thirdly, there is *The Design of Design* defining the underlying principles of turning an invention into a realistic machine. All are published by the Cambridge University Press.

Cambridge GORDON L. GLEGG
September 1973

1

THE OVERALL PICTURE—PLAN VIEW

Gazing at this sprawling subject we must begin with the problem of where to begin. Just to parachute down into it at some familiar spot and wander busily about will mean missing more than we see and dodging difficult spots. Aristotle said 'Those who wish to succeed must ask the right questions.'

Perhaps in this situation the right question is how to form an overall picture that includes all the variegated fields where research is essential as a foundation for engineering design.

This book is about something in particular, not about everything in general, and an overall picture should begin by showing us where the boundaries of the subject lie.

Elsewhere I have written 'The Engineering Scientist and the Natural Scientist travel the same road but sometimes in opposite directions. The Engineer goes from the abstract to the concrete: other Scientists from the concrete to the abstract. The Astronomer takes most careful and exact measurements of a planet and then deduces its future position and movements in the form of abstract mathematical formulae. The Engineer's work is the converse of this. He invents with his imagination and then builds with his hands...'[1]

In research for design the engineer and the scientist are combined into a single individual who does both jobs. He travels from one concrete reality to another via a horseshoe journey into the abstract, or, if you prefer it, from the real into the symbolic and out again into reality.

This, I suggest, is the overall picture for which we are looking and everything we shall consider prepares for it, exhibits it, or results from it. We shall analyse it in increasing detail but it will remain the same throughout.

Standing back and taking a general view of this picture or map of the land we are to explore we shall be immediately struck by some outstanding geographical features. The first of these is that all the roads end up in the same field from which they started. Despite long loops into the abstract they ultimately boomerang back to their home ground. The prodigal always returns. In short, you cannot safely base design in one field of engineering on research carried out in another.

This fact you might think to be so obvious as to be self-evident, but in practice losing sight of it is only too easy, and probably wastes more millions of pounds per year than any other type of design disaster. Unfortunately it is a danger that we are constantly exposed to for economic reasons. No one can afford to pay us to duplicate research already known to have been done. We must, perforce, accept other people's research results, but before we do so it is essential to identify the area in which they were carried out. And it is only too easy to forget to do this, for we are all conditioned to accept other people's figures uncritically, especially if they are written in books. This conditioning began, probably, at school. Rushing through our homework before our favourite television programme begins, we are faced with a problem about the weight of a liquid in an uncomfortably shaped bath. Well, that's easy, we'll look up the density of water, or whatever it's called, in the back of the book. Work it out to three places of decimals, that's bound to impress. Next question is about expansion – better keep one thumb in the back page where the tables are. And so it goes on; faced with highly improbable sounding materials behaving in highly improbable ways, we keep our mind a blank, look in the back of the book, and give the answer to eleven places of decimals to show our total grasp of the subject.

And when we first enter a design office we bring with us this 'look it up in the back of the book' habit deeply embedded. We never like to be far from some Engineer's Handbook and very useful they are too. And very dangerous. Dangerous in two ways that must be both appreciated and distinguished.

For instance, you look up a table giving the physical characteristics of commonly used engineering materials. In the first column, probably, you will see the figures for the ultimate strengths. Suppose the material you are interested in is shown

2

as having an ultimate strength of 70,000 lb/sq in (about 1,000 MN/m²). It is important to realize what this means. It means that the ultimate strength is never 70,000 lb/sq in. Never exactly that figure, with no margin of error whatever. And as your eyes travel across to other tabulated characteristics this marginal error increases explosively. If you read that, under fatigue strength, a certain number of reversals of load will be withstood by mild steel before failure, you can safely assume that, with good luck, the figure will not be more than a thousand per cent wrong; with normal luck, ten thousand per cent, and with bad luck, it's anybody's guess.[2] But the second danger is a subtler one. Fatigue strength, or creep[3] for that matter, depends basically on the shape and size of the test piece used in the experiment and the line of the load acting on it. By all means use the book's figures for designing a bridge or an aeroplane provided you construct both entirely of little bits of metal, identically shaped to the test pieces the book is referring to, and joined at their ends only through well-lubricated ball joints to keep the forces in line. And if you can't design your bridge or aeroplane like this you also cannot use the creep or fatigue figures that are only appropriate to these particular shapes. You must not drag figures sideways across a frontier from an alien field and expect them to be still reliable. When faced by the wreckage of a plane or a bridge, it is not a very convincing excuse to say that they must have read a different set of tables from the ones you did. In short, by all means use other people's results provided you do so intelligently, and this means identifying the area of reality within which the figures are relevant, and making sure that it shares a common ground with the one in which you are designing.

I do not want to give the impression that the blind and unthinking 'looking it up in the back of the book' is the only mistake one can make. There is another and even more dangerous one, i.e. looking things up in the front of the book; especially the chapter called 'Calculations for Engineers'.

Soon after the last war started, I, as a young man, with little more than a undergraduate's experience behind me, was catapulted with only two day's notice into a job with numerous responsibilities; all of them totally terrifying. Amongst them was the immediate responsibility for an existing research organization, the biggest of its type in the country, whose main directive was to discover means of saving steel in the design of reinforced concrete; a matter of great wartime urgency. My ignorance of reinforced concrete was profound but the staff stood hopefully around asking for instructions, my superiors were demanding quick information, and I had to speak up.

One of the research projects already on the books and on which an immediate start must be made was the problem of the design of steel stirrups to guard against shear failure in reinforced concrete beams, and especially the influence on it of stirrup steels of differing characteristics.

I hurriedly thought to myself that if I wanted to find out what happens during a shear failure I'd better design something that was going to fail in shear. If it did not try to fail by shear I would look as out of my depth as I was. I therefore designed a series of beams of different dimensions and with different stirrups to be tested to destruction in bending, and put in them quite an extraordinary amount of tensile and compressive reinforcement to make certain that the concrete would try to fail by shear first. We did our tests and there were shear failures each time and I gleefully anticipated being able to list in order of merit the materials and design combinations. And I couldn't. It was a chaotic self-contradictory jumble of figures. I was acutely embarrassed and wished I could sink through the heavily reinforced floor. After many hours of desperate thinking I found the explanation. The muddle was entirely my fault. In a simple-minded and uncritical way I had assumed that what I had read in a text book was true. 'The amount of tensile reinforcement in a beam has no influence on its shear strength', it had said. And, of course, this statement is quite true in the normal economically reinforced design. But I had designed in a different area, where the longitudinal reinforcement was artificially exaggerated to ensure shear failure. Why should a statement relevant to one specialized design area be necessarily true

4

of another? It needn't be and it wasn't. By assuming that the reinforcing bars had some real effect on shear strength it was fairly easy to assess a figure for it, and immediately all the results fell into a gratifyingly consistent pattern.

A parallel danger is one that, paradoxically, we are more inclined to run into the older we grow. We can be misled by our own automatic thinking.

In the course of time we become so familiar with certain common types of calculation that we begin to do them almost without thought at all. We have done the thinking so often before we feel we don't have to do it again. We have evolved a mental text book; pictures of principles and formulae that have become routine and which our memories can hand down to us at a moment's notice. All very nice and proper, until one day our memory hands down the wrong picture without us noticing it.

Imagine, for instance, that you are designing some big cooling towers for a power station. They are to be about 400 ft high and constructed of reinforced concrete. They have to withstand the forces of nature. The vital statistics of cooling towers are not very notable, but they do have an appreciable waist around the middle. This means that gravity tries to make them fall down outwards and wind pressure blow them down inwards. There is a continuous tug of war between the two forces. You must design the reinforcement to deal with the difference and, with high winds, it is gravity that loses. You make wind tunnel experiments and so estimate the maximum force that normal winds are likely to produce, subtract the outward force due to gravity and specify the reinforcement to deal with the difference, having added on a nice fat safety factor.

You build the towers and they almost immediately fall down again. And if you ask how do I know they would, my reply is that they *did* fall down; at Ferry Bridge in Yorkshire in November 1965.[4]

If you know little or nothing about reinforced concrete the fallacy in the preceding paragraph is probably easy to spot; not so easy if you are more experienced in the subject.

The normal technique in designing a reinforced structure is

5

first to find out all the forces that it must withstand. Its own weight, the applied loads and so on. As all these act in the same direction i.e. downwards, you add up the total, and work out the size of your reinforcement, incorporating the normal safety factor. And you do that so often that it becomes automatic. It is far from easy to pull yourself out of this mental groove and to realize that a cooling tower presents a different picture altogether. The forces no longer pull together but against each other. Gravity wants to pull it over outwards, wind inwards. Now the normal method of calculating would still be applicable if we could rely on gravity pulling harder every time the wind blew stronger. But it is very difficult to negotiate a local agreement with the forces of nature in this way. What actually happens is that the force due to gravity remains constant but as the wind pressure increases the out-of-balance difference between the two increases at a more rapid rate still. Your safety factor can easily be swallowed up; for it is safe when you add up loads, but not automatically so if applied to their differences. You should have put the safety margin on the wind load, not the reinforcement.

Beware then of being hypnotized by the familiar; keep both feet on the ground and be sure that it is the same soil that they are both on.

Lastly there is the possibility of failing to have our designs rooted in reality at all. We think they are, but in actual fact we are floating in the abstract air of the subjective. And this is the prime cause of the besetting sin of designers; the itch to 'improve' things. On the face of it, this seems a harmless and even meritorious aim. Actually it is generally disastrous.

For instance, a machine has been designed and is in the course of construction when suddenly the designer rushes up and insists on some alteration being made; he has thought of an 'improvement' which must be incorporated without delay. Drawings are hurriedly modified and new parts rushed through. And if the machine gives trouble after it is started up, the betting is that the last-minute 'improvement' is the cause.

Rightly or wrongly the U.S.A. has the reputation of being able to develop a new invention much more quickly than we do in this country. If this is true it may well be that one of the reasons for it is that the Americans usually veto any improvement in design after construction has begun. Leave it alone and alter the design in the next machine or the next batch; don't tinker with this one is their policy. And it is a highly realistic one.

Now at first sight it seems very odd that otherwise apparently normal individuals, having conscientiously based a design on their own research or the identifiable results of other people's, should lose their heads at the last moment and frantically demand modifications which turn out to be largely incompatible with the original design, or, at least, suffer from the unreliability of hastiness.

Actually there are two powerful forces triggering off this apparently insane behaviour. The first arises from the pattern of thought that the inventive mind often follows. I have written about this in detail elsewhere[5] but it can be summarized by saying that it is a combination of two mental attitudes. First comes the concentration on the problem and the deliberate saturation of the mind with all the aspects of it. Next, as an essential and conscious discipline, the inventor ceases all further thought on the problem, relaxes, plays a game or, if he must work, thinks about something quite different. This releases energy into his subconscious mind which works away at the problem and, after an interval, and often at the most unexpected moment, hands up a solution into the conscious mind where it is pictured in the imagination.

The inventor, hopping with joy, sets a drawing office to work and finally a machine is constructed. During this period, between invention and completion, which usually lasts months, sometimes years, the inventor cannot turn off his subconscious mind like a tap. As he switches to other work, or has a holiday, new ideas about the machine will suddenly occur to him, possibly better, possibly worse than the original one, and he will feel a compulsion to incorporate them. And naturally, time being short, these 'improvements' will not be able to have the experimental backing they ought. They will be like currants pushed

7

into a bun after it has been made and equally likely to fall out again.

Worse still, another subjective influence will multiply this unsettling feeling that a 'better' design is possible, and this effect is one that operates in many walks of life.

For instance, a general will plan how he intends fighting a battle and then must wait while all the military build-up takes place, and this waiting may be more trying than the battle. He has little to do and his original decisions churn round and round in his mind; he wonders if they were really the right ones and so on. If he is wise he will realize that this frame of mind, stirred up by energy bursting for action, is not only possible but largely inevitable, and sudden modifications to his original plans must be very carefully scrutinized to make sure that they have more than a mere subjective release value for a jumpy general.

In exactly the same way the itch to 'improve' a design is not only possible but almost inevitable in an engineer and, even if forewarned, is one most difficult to withstand. These two factors; the inventive that won't stop inventing and the tension of delay, pull together and may easily topple our objective reasoning, especially if we are not on guard against them.

When an undergraduate I spent my vacations inventing and building racing cars of original design and highly temperamental behaviour and I learnt much from the disasters that constantly occurred.

For instance, I entered a single-seater sprint hill climb car for the most important event of the year at Shelsley Walsh in Worcestershire. Lacking the finance for a transporter I had to tow the car there on the end of a rope. The stop–go traffic blocks were such that I had to use the brake almost continually to keep the rope tight. Unfortunately this only brake, which acted on the transmission, became hotter and hotter and finally disintegrated. I managed to pass the scrutinizing braking test by surreptitiously engaging reverse, but I had a feeling that a total absence of stopping power would add an extra hazard to what already promised to be an interesting afternoon. And then an even worse disaster occurred. It started to rain. A few weeks before the car had had its first outing at an event at Belfast where it had also rained. After one run I had been banned from taking further part as the combination of myself and the car was too dangerous for even the Irish to face a second time. And now it was raining again. I sat under an umbrella and waited for the start, a couple of hours away, and gloomily contemplated the grey skies and the almost totally smooth rear tyres of my car. A financial crisis had recently occurred as the result of purchasing new tyres for the front wheels and I was not in a fit state to buy new rear ones. Very kindly the garage had given me a pair free of charge; they were in advanced old age but reputed to hold air for a reasonable time.

As the minutes slowly ticked by I was gripped by an increasing feeling that I ought to do something about the situation. Finally I could stand it no longer, inaction was intolerable; I must try to improve the situation. I snatched up a saw and started cutting grooves into the woefully thin rubber on the rear wheels. Having done that I felt a good deal better, a feeling not shared by the tyres. A few minutes before I was due to leave the starting line one rear tyre gave a feeble explosion and went flat; I don't blame it. I jumped from the car, shouted to the starter that I would be ready in a moment, and bolted through the hedge which separated the pit area from the public car park. I looked wildly around, saw a car with the same centre lock hubs as mine and hurriedly removed its spare wheel. I bowled it back through the hedge and within seconds the waiting mechanics had it on, replacing the flat one. The fact that I finished first in my class was largely due to the lovely tread on the near side rear wheel. I then returned the wheel before the owner could get back to the car park, leaving a note thanking him for his kind and willing co-operation.

Ever since then I have been on my guard against hasty 'improvements' which are not rooted in identifiable realities and merely arise from the inevitable subjective feelings of a designer during periods of waiting, and I advise you to guard against them too. Know your enemy, especially if it is yourself.

2

THE OVERALL PICTURE—ELEVATION

Any design based partially or wholly in the abstract is a gamble. Concrete certainty lies only in research, but this costs money. Happy is the designer who discovers that all he needs has already been done and paid for by someone else. Otherwise he must do it himself and it will probably cost twice as much as he anticipates at the outset. Good research often costs a good deal of money, but spending a good deal of money does not automatically produce good research.

During the war I was given a research-design job of such priority that I could demand almost anything and get it instantly. But there was one condition, and one only, that I must fulfil. I must spend £156 a day. If I was not doing this the official attitude was that I was not working fast enough. So, each day, one first decided how to use up the money, and then got down to serious work. And this worship of the magic of money still survives, despite the fact that, more often, the reverse is true. In industry time is money; in research money, unwisely ladled out, means more time. Research on a shoestring would not take so long; there are less strings to it, and less possibilities of tangles. But there is a danger in going to the opposite extreme and arbitrarily reducing a research quota for economy reasons. It may be hopelessly uneconomic to do so. More than once I have been asked in to help a bewildered management swimming about in an abstract sea of opinions without being able to put even a toe on the concrete bottom, despite having spent money on 'research'. If research must be done, do it properly; otherwise don't do it at all. In this connection there are two pitfalls that you must watch out for. The first is that, having accurately identified the area in which your research must be done, you unwisely contract, to save money, the frontiers within which you work.

A few years ago many hundreds of thousands of pounds were

spent in erecting a factory and equipping it with a production line of a novel design. This new design was based on research carried out on an existing and quite successful one. It was discovered that an expensive part of the original process could be by-passed by an ingenious innovation, and a number of tests showed that it was a production possibility. I came on the scene after everything had been built but nothing was being produced.

The vital point about the novel part of the process was that the material was to be able to transport itself without support. 'We could do it easily on the old machine in our experimental runs' they said; and added that if the material showed any sign of collapse they could easily increase the amount of an ingredient in the material which I will call 'x' and this would give it added strength, at very little additional expense. And this was perfectly true; and it was equally true, as they subsequently discovered, that increasing 'x' made the material much more sticky and so needing extra strength to be able to transport itself. It had not occurred to anyone to do research into this possibility; and so they had ended up in double-crossing themselves for the stickiness increased more rapidly than the strength. They had chosen the right area for their research but limited its scope. You must never risk doing this. Remember that, in design variables, 'if it is possible, it is inevitable'. The possible sooner or later always happens. Speaking of the ill-fated airship R101, Lord Thompson of Cardington said 'She is safe as houses, except for the millionth chance' and a few hours later he was dead. In the area of research it is the frontier conditions that are often vital.

The second money-saving illusion is that of dotting your experiments about within the relevant area and then working these individual samples up into what you imagine to be a total picture of what is going on. The problem with this approach is the difficulty of deciding where to pick the spots for sampling. To be certain of that you would have to have a clear picture of what is going on, and if you knew that you would not need the experiments at all.

A large organization was having considerable trouble with a

production line making plastic materials where the finished appearance was vital, and this was influenced by a large variety of factors. The research department was working hard to discover the relationship between cause and effect while the management were stamping around impatiently demanding recommendations. Finally, and not unjustifiably, the management decided that they would by-pass the research department and do some 'practical' full scale experiments on the production line. They stopped normal production for a week and ran the machine, with all its attendant staff, under experimental conditions. They had worked out a detailed program in advance and the appalling expense would be justified by the achieving of some quick and realistic results, they said. They did the experiments, spent even more money than they had anticipated and produced a host of facts, figures and samples. These were analysed, quarrelled over, re-analysed, and finally set out in a report of considerable length and ambiguity. The management, hopping mad, asked me to review it all and submit a report stating exactly what had been found out. My verdict was 'nothing whatever'. Due to the interaction of variables that the production personnel had not anticipated, no logical relationship between the results could be established with any degree of certainty at all. All they learnt was that they had learnt nothing, except the extent of their own ignorance. Beware of spotty research. We will return to this point again later but meanwhile we must consider this problem of variables at a greater depth and the next three examples will give an introduction to what is meant by this.

Consider, for a moment, the design of a machine for hobbing gear teeth. You have made the frame of the machine so that it looks comfortingly rigid and arranged for it to be firmly bolted to a concrete floor. All should be as steady as a rock you reckon. But will it be? A concrete floor sits on earth; earth moves about if its water content varies; water comes in and out with the tides, and the tides depend on the moon. Are you sure the moon will not show its effect on the gear teeth? Your response to this fairy-tale sounding sequence of events is probably to call it a lot of moonshine. In actual fact there is a known case[6] where all

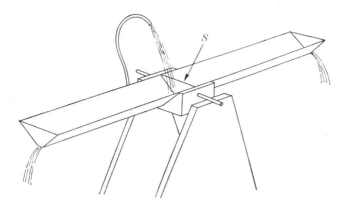

Fig. 1. The 'stabilized' seesaw, with septum S in position. The model was made by soldering two pieces of angle brass ($1\frac{1}{4} \times 1\frac{1}{4} \times \frac{3}{32}$ in), each 12 in long, on to a central brass block which had a 90° channel milled out of it; a gap of $\frac{1}{8}$ in was left between the ends so that a perspex septum, with bevelled upper edge, could be inserted. The ends of the channel were closed with plates and a no. 43 hole was drilled in the corner between channel and end-plates. The pivots were 8 B.A. steel screw threads covered with polythene sleeves and a counterweight (not shown) was attached beneath the central block so as to give a time of swing of about 6 seconds (this, like the other facts given here, is not at all critical). The water jet was drilled with a no. 43 drill, and a 1-foot head of water from a bottle, controlled a screw clip, gave an adequate flow. From *Reconciling Physics with Reality*, A. B. Pippard, C.U.P., 1972.

this actually happened and you could deduce the position of the tide when a gear had been cut in a seaside factory by examining the teeth afterwards.

Next we will look at the design of petrol engine camshafts and their attendant valves and actuating mechanisms. Using a computer it is not difficult to discover, for any given speed and design of cam, all the velocities, accelerations, forces and so on in all parts of the assembly and relate each to each. So far, so good, but it is not necessarily good enough. Your results are necessarily based on the assumption of certain fixed angular velocities for the cam. But there often aren't any. A camshaft is inherently torsionally unstable and therefore no two cams on it are going at the same angular velocity, except momentarily, and none of them ever achieves 'A fixed angular velocity' at all. With multi-cylinder engines and high r.p.m. this becomes a major problem.

13

The third example is that given in the fascinating inaugural lecture by Professor A. B. Pippard, F.R.S. on becoming Cavendish Professor of Physics at Cambridge.

'The toy seesaw shown in Fig. 1 rocks freely and I pretend that, wishing to stabilise it against disturbances due to chance breezes, I arrange that when one side tilts down the other side is automatically weighted to bring it back. This is achieved by fixing a septum in the middle and letting a trickle of water play on it (a small hole at each end allows the water to drain away). As soon as the seesaw tilts, all the water falls on the higher side and so realises my intentions. However, when I turn the water on, it is not many seconds before the seesaw starts swinging violently, and very soon it is tilting up to the vertical and even beyond, and will continue this as long as the water is kept running.

'Having demonstrated this behaviour, I remarked that perhaps I had made a mistake and that the water ought to run into the side that was lower, rather than higher; this is achieved by simply removing the septum. I asked the audience what would happen now, and soon those who were bold enough to speak had agreed among themselves that if it tilted down towards the left, it would continue tilting further until an equilibrium was reached with the water draining off as fast as it arrived. If it had tilted to the right, it would have stabilized itself tilted that way (Fig. 2). That was indeed my guess also when I first set the experiment up, and it is very nearly right – if one helps the seesaw to steady itself in a tipped position and then lets it go, it seems for some while as if it is quite stable; but gradually slight tremors begin to grow, until the rocking reaches an amplitude that brings the seesaw over the horizontal, the water rushes across and it begins its rocking motion round the alternative "equilibrium" position. The number of swings it makes on each side before tilting right over is highly variable (see Fig. 3).

'In Fig. 3 a simple recorder attached to the seesaw provides a record of its behaviour on a moving paper chart (angle of tilt plotted against time). In the upper diagram the septum is in place and it can be seen how rapidly the oscillations build up to

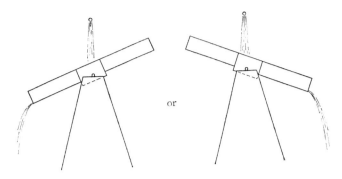

or

Fig. 2. The seesaw without septum, showing two possible positions of equilibrium. From *Reconciling Physics with Reality*.

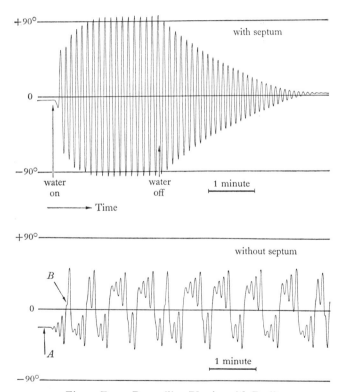

Fig. 3. From *Reconciling Physics with Reality*.

15

011 65

a more or less steady amplitude of more than 90° each way (the seesaw tilts beyond the vertical). When the water is turned off the oscillations decay gradually.

'In the lower diagram the septum is absent and after turning on the water the seesaw is encouraged to settle down in a stable state tilted down by about 20° (state A). It soon begins to rock, however, and at point B (which is far more of a cliffhanger than the emotionless chart reveals) it hesitates between swinging back to its original posture or plunging over to the opposite side; its choice of the latter alternative does not lead to a permanent change of side – indeed it immediately reverts. But later it spreads its favours equally between the two sides.'

We must now stand back a little and look at these last three examples and then we can find, I think, an evolving pattern of behaviour.

The machine tool was designed to hob teeth in steel. Everything was contrived to that end. A piece of metal never, by just being left around, succeeds in getting teeth like this cut into itself. The whole process is, in this sense, artificial and unnatural. But there is one large exterior factor that exerts a minor influence, i.e. the moon. Now the moon is in complete contrast to a man-made machine. The machine is artificial; the moon natural; we can move the machine about, or not use it, or sell it or knock it to bits, but we are stuck with the moon. It's an inherent part of our environment whether we want it or not.

Turning now to the camshaft we find there too the same contrasting pair of variables. We can alter our mechanisms about and modify our cam designs to our heart's content, but the torsional instability of a shaft is a natural phenomenon we cannot eliminate nor ignore with impunity. Here then the inherent variable may have a major rather than a marginal effect.

Turning finally to Professor Pippard's seesaw we find the natural characteristics have taken over almost completely from the artificial ones and have made a laughing stock of the original mental picture of how it was going to work. The stream of water, which we simple mindedly expected to be a force to stabilize a member is turned into a source of energy exciting a dynamic system. The inherent has almost totally engulfed the contrived.

This distinction between the artificial variables which are within our control and the natural ones which are not is one of kind and not degree, and so is most usefully represented by adding an extra dimension to our picture.

For instance, we will give to the highly independent movement of the moon and its effects on the tides the dimension of depth. Our contrived variables are thus an area, and our natural one a thickness, and our total picture is that of a thin disc.

Turning to the camshaft we find the depth increasing and we suddenly wake up to the fact that our overall picture is not that of a flat countryside bounded by frontiers but that of a flat topped mountain with sloping sides. We have to lop off quite a lot from the top of the original mountain so that we can level off at the depth of our natural variables and the total area of this new plateau is correspondingly increased. All research is a study of table mountains and the level at which we can safely cut them off. With the seesaw we must choose a special point almost at ground level and forget about our initial impression of a high and almost pointed apex.

Immediately arising from this is the question whether the choosing of special points and heights is not invalidating our original stipulation that all design must be looped back into the concrete. Can a 'special case' be considered homogeneous reality; have we not mixed a little abstract in with it? To help in finding an answer to this we must stand further back still, and this will enable us to see that the whole of our normal existence is on the level of a 'special case'. Above us is the vast-scaled world of general relativity, below the lawless minuteness of quantum physics. Humanity lives and moves and has its being on the wafer thin disc where these worlds infiltrate each other. We can climb up the mountain with Einstein and soon get giddy, we can go pot-holing with Rutherford and soon get lost, but at night we must return home to our special little ledge were we dwell with Newton; where we eat and sleep and even design things. To us the special level is normality; above or below are the special cases. To an onlooker we are perched on a freak frontier, we are the curiosities of the universe; we are the special case. And I wonder if we can stand further back still? Is it possible that the whole universe is itself a special case? Both the existence of matter as we know it and our experience of it is basically dependent on velocities of one kind or another, and the speed of light is the ultimate limit of any velocity in our universe. If anything went faster than this, we could not know of its existence, nor even imagine what it would be like. We may thus be the exception in a world of even greater reality, a world containing and infiltrating but not fettered to our time–space continuum, a world in which our universe is a curiosity; a special case.

In short, there seems to be a number of reasons for thinking that special cases are quite respectable and so we will now turn attention to the possibility of exploiting them for research purposes.

3

SPECIAL CASES

To be good at research you must be good at mountaineering. Scrambling up and down the sloping sides of the problem, you must find the highest level at which its essential features can still be preserved.

Imagine, for instance, that you have been unable to go and watch a special football match but you want to find out what happened. You rush home and begin doing research into the problem by pounding the buttons on your television set until you can find a film recording of the match. You then settle down to watch what happened. And you see what happened by not seeing what happened, not exactly that is.

Your television set only gives you a two-dimensional reproduction of the match; it has dropped a dimension. Now dropping this dimension has not altered the match: the same side wins, but it has enabled you to do the research you wanted on the apparatus you had available. It is special but not misleading. And this dimension-dropping gambit is perhaps the most obvious way to simplify research, provided we do not forget to put it back when it is needed.

For instance, if you are designing a space vehicle, you must study the influence that other distant space objects may have on its flight. For the sake of simplification astronomers assume in calculating these gravitational forces that the total mass of the star or planet is concentrated within a point of no magnitude at its centre. This is a useful gambit but if you forget to put the missing dimensions back again there will be a nasty and complicated bump when the space vehicle tries to land on anything.

But linear dimensions are not the only ones we can rise above; we can make time stand still and, with it, velocity and acceleration which have no separate existence.

For instance we may want to see if the ball did really cross the goal line and so the television plays back a slow motion edition of the shot, finally stopping the film to give us a 'still' picture of the decisive moment.

I was bull-dozed into discovering this technique in the reinforced concrete laboratory that I have already mentioned. At the same time as investigating the shear failures I was instructed to consider the design of air-raid shelters and in particular the Government regulations concerning their construction.

This code of practice stipulated that the relative working stresses in the different type of commercially available reinforcing steels were to be in the same ratio as that laid down for dealing with conventional static loads. In those days my views were less defined than they are now, but I instinctively rebelled against the suggestion that what was true for static loads was automatically true for dynamic ones. I could imagine that the static could be a special case of the dynamic; leaving out the time dimension might do it, but not the other way round. Moreover the code of practice for normal loads was based solely on the yield point of the steels, a figure which appeared to me to be actually misleading as a criterion of their energy absorbing properties.

But this was not all. Why assume that the overall behaviour of a structure would be the same whether a piano was gently put on it or a bomb hurriedly dropped on it? In short, I felt that the Government Department concerned had left out two vital dimensions, apparently without even realizing that they were there. I sat down and tried to think out a viable solution and I soon realized that, having scrambled down two dimensions with relish to realism, current industrial considerations would force me to climb laboriously back up again. For testing high speed loading, a high speed testing machine should be used. Neither I nor anyone else had one. To specify materials in such a way that no one could do the test wasn't going to help anyone very much.

Secondly, and equally frustrating, was that even on a slow loading machine I would still be compelled to limit myself to existing test procedures as the introduction of new ones would meet with massive opposition and delay. I began to sympathize with the Government Department that decided to let well, or rather bad, alone. I was reluctantly convinced that the only acceptable solution was to fabricate up from conventional static loading tests a special case which would accurately correspond to the deeper realism of the problem. I therefore obtained the appropriate British standard specification and a copy of the code of practice, and studied them. For reinforcing steels, and there were three different types, the B.S.S. specified, for each, three tests while the code of practice only took any notice of the results of one of them. This, to a beginner, seemed a bit odd. The code of practice based the working stress in steel reinforcement on its minimum yield or proof stress and nothing else. The corresponding B.S.S. gave the acceptable minimum values for this in mild steel, hard drawn wire, and strain hardened steel, and then went on to specify minimum figures for ductility and ultimate strength.

As the code of practice not only ignored these figures but ignored their existence, I formed the conclusion that they were included solely as an indirect guide to the quality of the steel. The presence of impurities in the metal would be quickly reflected in the ductility or ultimate strength, while the yield point might easily remain unaffected. Thus I had available, luckily, three commonly accepted tests that I could juggle about with. Obviously I must study in more detail what these tests portrayed and so I examined typical stress–strain diagrams as in Fig. 4. If one assumed, for the moment, that the energy absorbed by a reinforced concrete beam was largely dependent on the reinforcement and that the area under its stress–strain diagram was a guide to this, the types of steel must vary enormously in their abilities to cope with such loads. Hard drawn wire's area was less than a quarter that of mild steel. Were these results dependent in any way on the method of testing was the next question to be asked? The ultimate strength seemed unlikely to be affected but the ductility was a more complicated matter. Before placing the steel sample in the tensile testing machine a number of equally spaced marks were made on it. After fracture the final ductility was defined as the

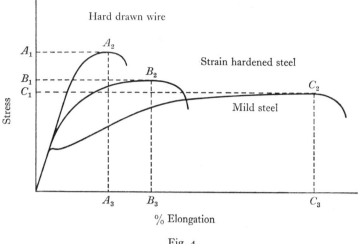

Fig. 4.

extension in spacing between the two points that included the fracture, measured by putting together again the two halves of the specimen. Therefore the percentage ductility was certainly dependent to a certain extent on the testing procedure; if you used wider spacing for your markings, or if you measured the extension between two marks that did not include the fracture, your results would be numerically different.

This can be seen by examining the stress–strain diagram again. The shapes of the three curves are unaffected by the distance over which you measure the percentage elongation, up to the point of maximum stress. The shape of the curves after this is almost totally dependent on how you do your measuring. If you confine your elongation measurement to a pair of points that do not include that where the steel 'necks' and fails, your stress–strain curve will fall vertically down to zero immediately after passing the maximum stress. With the failure located between two points, the extension of the curve after its maximum height depends almost entirely on their spacing.

As the effective length over which energy is absorbed in reinforcing steel in a beam is certainly going to be greater than the small and arbitrary test length, we should only include, in defining dynamic capabilities, the extension at maximum load. But this, in itself, is a new definition and therefore inadmissable; can we get at it some other way? Very approximately, the downward sloping part of the stress–strain diagram occupies about five per cent of the elongation scale in all three steels. The next point that occurred to me was that the area under the stress–strain curve was, approximately, a constant proportion of the area of the smallest rectangle that included the maximum stress, i.e. A_1, A_2, A_3, B_1, B_2, B_3, C_1, C_2 C_3. In short the relative energy absorbing capabilities of steels could be determined by multiplying their ultimate failing stress by their formally defined ductility less 5%, using B.S.S. test procedures.

Now all this was pure conjecture, fabricated up from a consideration of the stress–strain characteristics and the necessity of employing already familiar definitions. There was only one way of discovering if this hypothesis was grounded in any degree of reality and that was to test actual beams with energy loads. The beams were built: large weights were dropped from the top of the roof, a hazardous and noisy form of gymnastics, which was nearly assisted by a real bomb coming down on top of it all, and the failing loads of the beams were found to reproduce my theoretical analysis to a surprising degree of accuracy. Overjoyed I rushed off to the appropriate authorities and demanded that their regulations should be re-written to conform to my findings. They took not the slightest notice. Being young, and, I suspect, not a little uppety, I was not easily sat on, and reacted by writing to the top man in the country, Professor Sir John Baker, Scientific Adviser to the Government. Even at that age I had learnt that it is better to go to the top and work down rather than go to the bottom and work up. Sir John was in a position to thump Government Departments, it took him ten minutes to make up his mind about my results, and he immediately delivered the necessary thump, with gratifying results.

This dimension-dropping technique has a wider variety of applications than you might at first imagine. We will go back, for an example, to the television set where you are still watching the football match. You observe, with gratification, that when the referee is not looking the centre forward delivers a smart uppercut to the opposing back who measures his length on the ground. Undoubtedly a powerful blow you must admit. But why must you admit it? A television set cannot transmit or convey forces. You may throw something at it, but it is unlikely to throw anything back at you. It has no dimension for force. The full back felt the blow, you observe a deflection, for him the force is an experience, for you a deduction. 'A good hefty punch' you say, 'look how far he was hit'. You extended the use of the dimension, (distance) to take the place of another (force).

An example of this is a method of research into the relationship of load and deflection in complicated structures. Uncertain about what may happen, you build a prototype for testing and are preparing to load it and measure deflections at appropriate points. And then you realize that you may well be able to drop the load dimension altogether by the use of 'influence lines'[7]. This gambit exploits the fact that if you artificially deflect a structure at a certain point the remainder distorts slightly so as to absorb the effect. Make a map of this now distorted framework and so discover the influence that the artificial displacement has on any given point in it. Working back from this you can now calculate the effect of the applied force. As an experimental dimension direct load measurements can be eliminated in this way. And in process engineering you can use the same gambit. In Fig. 5 there are three pairs of rolls mangling out a plastic material of a rather temperamental type. This is a simplified version of a typical process problem where the finished thickness and appearance of the sheet is vital and very susceptible to the rolling conditions. You are asked to design a new machine to do the job faster and better. Wisely you are determined to base your new design on concrete facts and figures derived from the existing one. You are also convinced that the only way of doing this effectively is to alter one adjustment at a time and observe its effect, thus building up an unmistakable and unambiguous picture of what is going on. First you mount suitable instruments that give you the roll pressures p_1, p_2, p_3, the initial and

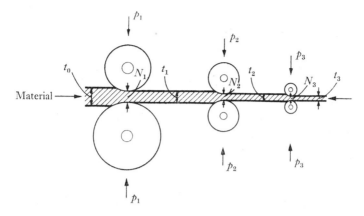

Fig. 5

subsequent thickness t_0, t_1, t_2, t_3, and widths in the nips of the rolls, N_1, N_2, N_3.

You have been told that the roll pressures are the most influential factor and so you decide to start off with varying them. You take out your notebook, increase p_1 a small amount, and watch what happens. You find that p_2 and p_3 immediately alter their values in sympathy, so have t_1, t_2 and t_3. Well, you are determined to alter only one variable at a time so you feel you had better put p_2 back to its original figure. This again instantly affects the readings for t_2, t_3 and p_3. What do I now do with p_3 you wonder? Put it back to its original value too? But that produces a most undesirable effect on the appearance of the sheet and its thickness becomes far outside production limits. You gaze at a blank notebook with a seething mind. What can you do now? You were determined on a logical approach with only one value altering at a time and you find yourself sunk without a trace beneath a hopeless flood of interconnected and uncooperative variables.

This depressing situation has arisen because you were in too much of a hurry; now sit back, survey the scene and think. You have been told that the pressure the rolls exert on the material is vital, but it isn't true. The rollers do not exert pressure on it, the material exerts the pressure on the rolls; no material; no pressure. The pressures are a function of t_0 and N_1, t_1 and N_2, and t_2 and N_3 respectively. In fact you can drop the pressure dimension out of the experiments entirely and in so doing make possible your initial intention of only altering one thing at a time. For instance, if you decrease N_1, it will have no effect on N_2 or N_3, and so you can measure t_1, t_2, t_3, and know that they are the sole and direct result of adjusting N_1. Now you can proceed with further adjustments. It may be difficult, it may be very difficult, but it will not be hopeless as it was when p_1, p_2, p_3 were putting their oars in.

The next technique for simplifying research, having exploited dimension dropping, is to try dimension swopping. We will go back to the football match where the back, in a furious temper, is glowering at the referee and muttering to himself. Next time the ball comes near, he kicks it a tremendous distance and it ends up in the back of the stands.

The television camera follows the flight of the ball with relish and it is useful to note how it does this. The camera is moved so to keep the ball in the centre of the picture. As a result the ball appears to be standing still and the crowd is flashing by. We have totally swopped their velocities. This strategy can be exploited in a number of ways; one of which is that for testing tyres. A car travels along a bumpy road, its wheels travel with it, they also move up and down with bumps and revolve relative to the car. We want to see how variations in tyre design affect the movement of the wheels and of the car. Of course we can rush along beside the car with cameras and instruments but it is not a very attractive idea. It is better to swop two of the velocities; make the car stand still and the road move. This can be done by simulating a bumpy road by a slat-covered drum which revolves at varying speeds with the wheel and its tyre resting on it. The axis of the wheel is now stationary and instrumentation is much easier and the slats can be reshaped or repositioned to represent different types of road.

Wind tunnels are another example. An aircraft normally flies through the air; well we'll let the air fly past the aircraft for a change. And it is only a change, if you do it properly, in the

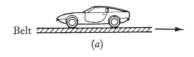

Belt

(a)

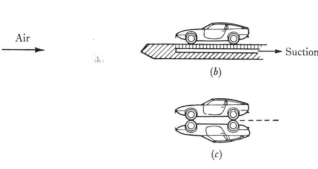

Air

Suction

(b)

(c)

Fig. 6

sense that the velocities are exchanged. Reality is unaffected; you are still watching the same football match but the observation of it has been made easier.

I said earlier that this procedure consists in the 'total' exchange of dimensions, and this word is vital.

You can do it for an aircraft in a wind tunnel but what about testing for the drag coefficient for a car in the same way? This is where you must watch out. A car moves through stationary air and along a stationary road. If you swop the velocities it must be a total one; you cannot leave the road out of it. If you have a stationary car and moving air you must have a moving road too. If you do not, you will find that in the wind tunnel there is a boundary layer of slower air close to the 'road' which is the odd man out. Various ways have been evolved for compensating for it,[8] none completely successful, and some of them are illustrated in Fig. 6.

(a) uses a moving belt to simulate the road, but it is very difficult to mount the car without disturbing the air.

(b) sucks the air away, hoping to neutralize the boundary layer.

27

(c) dispenses with the 'road' altogether by using a pair of cars, one inverted, with their wheels touching. Expensive and a bit unpredictable, it at least banishes the boundary layer.

But if all else fails and you cannot drop a dimension or swop it, you can try standing it on its head by giving it a negative value, and so using it as a ladder to climb up into a simpler climate. You can run the film of the football match through backwards and the ball will shoot out of the goal instead of into it.

For instance, if you are bogged down in endless process complications try doing your research on the assumption that a machine runs in the opposite direction to the normal.

I was once faced with a complicated and lengthy machine which was fed from silos, with five vastly different materials at one end and produced finished plastic sheeting at the other, and I was asked to do some research to find a relationship between what went in and what came out, with a possibility o redesigning the machine. For reasons that I have discussed more fully elsewhere,[9] most process machines buy their raw materials by weight and sell their finished products by volume and this was no exception. Not unexpectedly, the cost per unit weight of the five raw materials varied over a wide range. As each had a varying and unknown amount of entrained air, generally quite an amount of it, there was no known relationship between cost by weight and cost by volume, nor could one be established. When the five materials had been mixed, processed and finally squeezed into their saleable shape most, but not quite all, of the entrained air had escaped, thus introducing another variable. Finally, within the finished sheet of material, there were five different raw materials mixed up together in entirely unpredictable proportions by volume and even more obscure proportions by cost. For example, the substitution of a more expensive material for a cheaper one could well prove an economic asset if the former carried more entrained air with it through the process and, by reducing the overall density of the final sheet, eked out further all the other materials too. The only known way of ascertaining the economic effect of changing the ratios or natures of the raw materials was by doing a controlled pro-

28

duction run where the total initial cost could be related to the corresponding total volume of the final products. It was all most expensive, unpredictable, inconvenient, and often inconclusive. I found that the solution lay in imagining what would happen if the machine ran backwards. If this happened the 'raw material' would be the finished product which would then be pulled back into the finishing rolls, sucked back through the machine, split into its separate materials, air added, then shot back into the storage silos.

Now let's have a good look at our 'raw material', a compressed mixture of five elements. They are mixed up together but their total volume and mass would be unaffected if we regarded them as unmixed, a five-ply material so to speak. But each of these five plys will have withstood the same vertical and processing pressures so we can consider each in turn as if it was the only one. Knowing these process forces we can build up a laboratory rig to reproduce them and so establish the actual final density of each material in turn. As you know the cost per unit weight you can translate this into a cost-density. An example may help. Suppose you use as a cost comparison the value of the materials in square metre of finished sheeting, 1 mm thick. Having found your effective density and so the cost per unit volume you could make a list of your materials which might be something like this, using imaginary figures.

Material	Cost of 1 square metre, 1 mm thick
A	100 p
B	80 p
C	60 p
D	40 p
E	20 p

If there was an equal amount of each material present, each could be regarded as being in a sheet 1 mm thick, and the total cost of the five would be the total of all of them, i.e. 300 p. If you wanted to discover what the effect would be of halving the quantity of '*A*', and increasing, say, *D* and *E* by a quarter to keep the total thickness constant, you would subtract 50, add 20p and 10p and the new total cost would be 280p.

How would you ensure that the correct proportions of materials were fed into the machine initially to end up with the right proportions? That is quite easy. By imagining the machine in reverse we visualize the five materials filling up their silos, and the rate at which they do so depends on the speed of the machine. The five silos will steadily gain in the value of their contents and the relative rates at which they do so will be the same as the relative cost of them in the finished sheet. As we know the cost per unit weight, we can easily fix the ratios of comparative weights that must be fed in when we start the machine going in its normal direction again. Once you are used to the idea the arithmetic is ridiculously simple. You can turn up at a Board Meeting and with a few quick calculations, often done in your head, you can state what the effect on cost will be of alterations to materials or their proportions. This is regarded as some kind of conjuring trick and everyone is suitably impressed, provided you don't tell them how simple it really is.

Everything so far in this chapter assumes that the designer is in control of events. But can we be sure of this? Having settled down for a quiet picnic on a plateau with an attractive looking dimension, he may be rudely interrupted by the arrival of an embarrassing interloper. In dimensions, as elsewhere, two is company, three is none. Is there any way of insuring against this hazard? Yes, I think there are at least four and the time has come to discuss them in detail.

4

A ROGUE OF VARIABLES

If you ask a footballer what are the dimensions of the goal he will probably say something rude. The effect of his remark may well be that he is more concerned in getting the ball into it than wearing himself out remembering its measurements. He may add that as long as both goals are the same standard size that is all that matters; the knowledge of its numerical definition does not help you to kick a goal.

Similarly, in research, common dimensions are fair to both sides and cancel out; you do not need to know their absolute values.

I was once concerned in the design of a new type of centrifugal pump for use in paper mills. Paper is made of pulp and this is transported about the factory by mixing it with differing amounts of water as it goes through the various processes. In all paper mills, and especially in old ones, the number of centrifugal pumps to be found is astronomical and, even worse, hardly any two of them are of the same size. The design of each has not only to be matched to the widely varying ratio of pulp to water, which sometimes attains a porridge-like consistency and at others that of watered down milk, but also must be tuned to the local requirements of head and quantity per minute. The maintenance department's store is packed with a huge range of spare motors, pump casing, impellers etc., but hardly ever has the one you want if needed quickly. And the capital expenditure locked up in these spares is considerable and, if released, a more profitable job could be easily found for it.

It occurred to me that a possible solution to this problem might lie in having a universal pump that could be used almost anywhere in the mill. It would have to be able to deal with the different pulp densities without discrimination and must also absorb substantially the same amount of power independent of the head against which it worked. This means that at low heads a self-throttling of quantity would be an essential characteristic to stop the power going off the map, and, finally, the overall efficiency under all conditions must still be within commercial practice.

A difficult job perhaps, but the reward of cutting down the types of spares to less than five per cent of the normal figure was an attractive one. Although mathematics and hydraulic technology could help a little, much of the design had to be based on research. And the research had to be done within some standard conditions and there weren't any. It was impossible to define in any absolute way the turbulent muddle of all the pump's interacting requirements.

Well, if you can't beat them, join them. If you can't define a thing don't try. So I built two large tanks on a suitable erection and ordered a large quantity of high density pulp from a local paper mill. To compare the performance and efficiency of my research pumps with those of orthodox design, I used each to pump, in turn, the pulp contents of one tank into the other via a restricted orifice which could be adjusted to give the required back-pressure to simulate head, and if I wanted to alter the pulp consistency I merely added or filtered off water. I could never define in absolute terms what the cellulose content of the pulp was, nor its viscosity, flow properties, density range etc., but I did know that the tests were equally fair for each pump. The absolute remained veiled in mystery, but the relative results could be strictly compared. This type of testing is very well known and is generally defined as using a 'control'. I have never understood why. It is a basis of comparison; it does not control anything.

In short, one way of immobilizing disturbing rogue dimensions is to lock them all up together and only negotiate with them *en masse*.

Meanwhile, back at the television set, you are still observing the match. Normally this does not necessarily involve observing the crowd. But, all the same, you know that a crowd is there, for you can hear their cheering, an unmistakable noise.

Now this predictable sound is the cumulative effect of a large number of people shouting their heads off. No two of them will have identical voices and no single voice will, in fact, be making a sound exactly like that of the cheering. Everything is random and spontaneous. But there is more to it than that. Each individual voice is a multiple mixture of sounds and harmonics, almost infinitely complicated. Average out all these unpredictable sounds and you produce a predictable noise. *En masse* they are predictable precisely because they are individually unpredictable. You can only, with certainty, know the unknown. The absolute is the child of random parents.

In other realms of the physical world the same principle applies. If you say that a weight in a vacuum falls with an absolute acceleration of g, you mean that none of it does. None of the individual atoms, each in frantic motion, has a movement that coincides with the overall average of all of them. But we can go further. The paths of the unimaginable particles of energy forming an atom are also random and unpredictable. In the falling weight the unpredictable is piled up layers deep and again it builds the absolute. We can define all acceleration in the terms of g because it is an absolute, and it has become one because all possible variables are already included. We are down at rock bottom; there is nothing deeper out of which a rogue dimension can climb and plague us.

In research it is vitally important to know how to distinguish between the comparative and the absolute and so exploit each appropriately. Imagine two cars travelling along a road; and that their speeds differ by 10 m.p.h. (about 16 km/h). So far so good, but we can in no way deduce the absolute speed of either of them. However, tell us the speed of one of them and we can compare the other with it and so know the absolute speed of both. Compare something with a variable and you find a comparison, compare it with an absolute and you can find another absolute.

When I was testing those centrifugal pumps their performances were comparative with each other but absolute in the measurements that were compared. The pulp was indeterminate but power, angular velocity, linear distances, were all definite. Don't, I implore you, muddle the two things up.

I was once asked to report on a tricky situation where a new and complicated electro-magnetic chemical process was reputed to give great economic advantages. Well, at least, that was what the works chemical department said, but, unfortunately an august Government investigating team said that it did more harm than good. And my verdict hinged on one discovery, i.e. the works chemists had dressed up a comparative figure to look like an absolute one. But it was a long time before they would admit it! Almost certainly, when faced with the problem of planning research to fill up gaps in data for design your first decision will involve a choice between the comparative and the absolute.

At first sight, the comparative may well appear to be the most attractive. There are less things to measure, less things to measure wrong, and fewer expensive instruments to do it with. But, unhappily, everything is not always as simple as this. In practice, we are faced with various combinations of the two as well, and all have their particular liabilities and assets.

(i) The comparative-comparative

Imagine that you are concerned with the design of a big-end bearing for a car's connecting rod. You have drawn out a new type, better supported perhaps than the older one, and you need experimental data to confirm (or otherwise) your ideas. You approach the local car hire firm and arrange to put your new bearings in ten of his cars. After they have each gone 10,000 miles he is to return the cars to you, along with ten of his usual fleet, and you will take them to bits and look at the bearings. After this inspection, you will arrange them in order of healthiness, all eighty of them (assuming four cylinders in each car). If your new bearings are largely located among the top forty you will be justifiably pleased. Notice that neither the tests nor the results have any absolute standard or definition, both are strictly comparative.

This type of experiment is cheap and simple but has two main liabilities.

The first is that although you obtain your end in view, it is, in a sense, a dead end. Nothing else can be deduced from it; you can know nothing more than you already know. Apart from the one fact that is left over when the unknowns have cancelled each other out there is nothing left; there can't be. You cannot, for instance, forecast exactly how your bearing will behave under any other set of circumstances.

And the other difficulty is that of being certain whether your two comparative sets of cars were truly comparative, for you have no absolute way of checking it. In some of the cars, unknown to you, the random variables may have been modulated by an intrusive variable, and this may have unfairly affected the results. The normal ones cancel out, but abnormal ones may not.

Returning once again to the football match where you sometimes hear cheering: you may also hear something else, something quite horrible. It is called singing. To egg on their flagging side, the supporters burst into song. Their random mixture of voices, some deep, some squeaky, which, left to themselves, would average out into a consistent cheer, are now trying to conform to an applied variable and so are no longer totally random.

Are you sure all your cars were free from any abnormal influences? You say you are, but you can't really tell; you have no absolute way of checking up. Suppose, for instance, that most of the cars with the standard bearings had been hired out, under contract, to harassed wives with small children. Each morning the mob would have to be taken to school. With two children, a tennis racket, two satchels and the baby squeezed in the front beside the mother, she cannot get at the gear lever and, in any case, her left hand is heavily occupied in maintaining martial law in the back seats. As a result, for naturally they are always running late, the car is driven at a high speed in a low gear with a cold engine and the choke out. The big-end bearings suffer acutely. Can you be certain that the comparative advantage of your new bearings is not largely or solely due to fortunate absence of maddened mothers?

Summarizing, we can say, in general, that when we have a research procedure that uses comparative experiments and the results are also assessed solely by comparison amongst themselves we achieve one considerable asset at the cost of two liabilities. Our asset is simplicity; complications cancel. The liabilities are the possibility of abnormal influences affecting our conclusions and a total absence of absolute values.

We cannot safeguard ourselves against the first of these by inspecting the bearings themselves. If two cars are supposed to have a relative speed of 5 m.p.h. but do not, we cannot tell solely from a knowledge of the new relative speed which one is out of step; perhaps both are. Similarly, in our bearings, there can be no internal unambiguous evidence of exceptional exterior influences. The best thing we can do is to check as far as we can by thinking of a possibility and then making sure it is not there.

The second liability, the absence of the absolute, means that we cannot transpose our findings into any other context. We cannot deduce, with certainty, how the bearings would behave if fitted to an engine of rather different design. This leads us logically to the next combination, i.e.

(ii) The absolute-comparative

For this we build up a research rig that can apply known values for loading our two types of big-end bearings and then compare the results as before.

The rig may well be costly and complicated, for we must simulate the conditions inside a crankcase, especially with regard to lubrication and temperature, and then run each type of bearing for so many hours at various revs-per-minute. In this way we can hedge out the abnormal and so deal with the first liability of the previous method. Moreover, as our tests can be adjusted to absolute values of loading we can vary them to produce conditions to be expected in other designs of engines. And although our assessment of the results is still comparative, it is comparative in a rather different sense. We can now say that this is how a particular bearing looks after a certain running period of known severity, and this is how another one looks. You are comparing the two by evaluating their reactions to certain absolute conditions. And this means that you can test a bearing by itself; a rival design is unnecessary. The absolute-comparative can stand in its own grounds. This technique is that commonly used in ecology, in fact it has to be.

Imagine you are carrying out research into the design of an office chair. The chair itself can be defined by its shape and dimensions in absolute terms but not its comfort. You can assess that only in two ways, both comparative. You can either say 'this design is more comfortable than that' or 'this design is comfortable, or uncomfortable'. The comparison is therefore either between two degrees of comfort or between a particular degree of comfort and your subjective feeling of what comfort should be, a feeling that is impossible to define in an absolute way and will, in any case, vary with your degree of fatigue. In these circumstances the 'absolute-comparative' is the best we can do.

(iii)   The comparative-absolute

This opens the door to a new technique hitherto closed to us. For the ability to define results in absolute terms enables us to exploit the science of statistics. These most useful methods of analysis have applications across the whole realm of scientific research and are not the particular property of the designer, and so it is not necessary to expound them here where we are confining ourselves to the particular features of the research-design world. Nor, in any case, would we have the space to do so. We can borrow such techniques from other realms and other books[10] and apply them to our own ends, and so we need discuss them here only in a general way.

The first point to note is that statistics is the mathematics of absolute values. It enables us to analyse research and extract information otherwise hidden from view, but only if the individual results are presented in numerical form, in absolute figures.

The second point is that it enables us to detect abnormalities whose possible existence always haunts results only definable in comparative terms.

My experiments in the constant-power pulp pumps were of

the comparative-absolute type as power, head quantity and so on could be expressed in independent numerals. And so statistics could be a watch-dog. If the results were being gradually affected by the beating effect on the pulp produced by the revolving pump impellers it could be easily discovered as a drift with time.

The third feature of statistics, and possibly the most useful of all, is that it can often make sense out of apparently non-sensical results.

Reverting to our big-end bearings, it would be most annoying and frustrating if when arranged in order of comparative healthiness the new and old designs presented no obvious pattern.

We could guess anything but be sure of nothing. The solution would lie in abandoning the comparative and giving absolute figures for the wear in each bearing. You would measure them before and after. You could then discover the range of variation of wear in the same design of bearings in the same engine and go on to analyse in increasing depth. You would have every chance of deducing the presence of the harassed mothers in principle if not in detail.

In addition to allowing in the filter of statistics, ending with an 'absolute' has two further advantages. We may be able, for instance, to lift our results into higher dimensions. We might assess the rate of wear as well as the incidence of wear in our bearings. Not only statistical mathematics but all other manipulations are potentially exploitable. You can analyse the absolute in many ways, the comparative in none, and this leads us to the final advantage, that of being able to feed our results straight into a digital computer,[11] thus saving much both in time and human error.

(iv)   The absolute-absolute

The final double act is the absolute-absolute. We have already seen that, in general, an 'absolute' has a much greater potential value as a research tool than a corresponding 'comparative'. It is also likely to cost much more. Put two absolutes together and you find that their good characteristics add up, and so does their expense. Always choose the double absolute every time, if you can afford it. Alarming as the initial cost may appear there are at least three ways in which it may drastically reduce costs in the long run, but this only applies to a pair of absolutes; never to one only.

The allying of a comparative with an absolute is a treaty of friendship between two foreign countries; there is a distinct frontier where you pass from one to the other. They are strangers to each other, with different and possibly incompatible customs and language. The absolute talks in mathematics, the comparative in prose where the words 'better' or 'worse' constantly recur. It is a partnership where misunderstandings can easily arise. In contrast, the absolute-absolute share a common language and are basically compatible. You can pass from one to the other without being aware of crossing any definite frontier. Alternatively, we may picture it as the marriage of an understanding couple with common interests; and, as often happens, they start a family. After a time, a little absolute arrives sharing equally the characteristics of both parents. And his arrival is, unlike the normal procedure, the signal for saving money.

Having carried out research where the whole procedure is defined and controlled in absolute terms, there is always a chance that, between our experiments and their results, we may find a mathematical relationship. This would mean that we can dispense with further research and just calculate in absolute terms what the results would be. You might well start with only a little theory, but it could grow up into a big one. The vast literature on the 'mechanics of machines'[12] or the 'theory of structures'[13] all had little beginnings as the result of a couple of absolutes getting along well with each other, but they have grown steadily ever since. You can appreciate the advantages of such infant prodigies whenever you go out in a car.

For example, there are two distinct methods of finding out how best to balance a 60° vee-four internal combustion engine and so avoid shaking the passengers about. The first way is to build up a crankshaft and piston assembly complying to the absolute dimensions you have picked upon for the stroke, bearing diameters, and so on, and then mount it all in a rig with facilities for rotating it at various speeds and with numerous measuring instruments to record the exact vibrations arising in the crankcase and cylinder block. You fasten on balancing weights with definite dimensions at various points and record the vibrations in absolute terms. To reach a reasonable result is a very long job but it can be done in time. The second method is to avoid the expense of doing any research at all. You just calculate the answers. From research and analysis of balancing problems in the past a comprehensive mathematical picture has been built up and if you exploit this you not only save all the time and money but also have a distinct chance of arriving at a more precise solution in any case. A good way to balance such an engine, for instance, might be to have a separate balancing shaft with weights rotating in the opposite direction to the crankshaft. You would have to have done a great deal of *ad hoc* testing before you came up with that one.

But it is not only the roughness of the engine that irritates passengers, the roughness of the road does too. As we have already noted discomfort is an ecological problem that cannot be absolutely defined. And neither can the roughness of the road. You can say one road is more bumpy than another but that is all. There is no totally objective and absolute definition of the average road or the normal variations from it. The best we can do is to try out various suspension systems on various roads and choose which seems most comfortable on the whole. Everything is comparative, nothing absolute. And comparatives don't have families. No fundamental mathematical theory has yet arrived but a great deal of work is being done to discover one. Meanwhile we are stuck with comparative experiments and bumpy cars. Summarizing, we can say that the comparative-comparative gives a particular relationship, but the expensive absolute-absolute may well produce a general principle. You must sometimes spend money to save it. But you can also save money by saving it, and this possibility we will next discuss.

5

JOURNEYS INTO LILLIPUT

You can often save money by doing your research on a model of the real problem, though this is not necessarily as simple as it sounds. In making a scaled-up or a scaled-down model you cannot always assume that every dimension can be blindly modified in the same fixed ratio.

Suppose you are still watching your television football match and are shown a distant view to enable you to see the whole field and the progress of the play. The players look like dwarfs. But how can you be certain that they aren't dwarfs? Perhaps the manager of the match has decided to save capital outlay by reducing the size of the pitch and has trained a number of dwarfs to play the game; as they might drink less he wouldn't have to pay them so much. How do you know, merely from watching your television screen, that you are not seeing a consistently scaled-down version of the match? You will probably be able to guess by looking at how the players behave. Soon after the beginning of the match they may well become more interested in kicking each other than in kicking the ball. Why doesn't the referee blow his whistle and send someone off? But the wretched little man is blowing into his whistle with all his might, to no effect. A scaled-down whistle will either produce an unrecognizable squeak or else a sound above the human range of detectable frequencies. The dogs in the crowd would be most interested; not so the dwarfs. It can therefore be seen that the scaling down process has fundamentally modified our football match, the other side might now win. We have lost our realism.

It follows that if we want reality reproduced on a different scale we cannot just push it sideways across a frontier into a Lilliputian world where everything is scaled down uniformly. We must go back to the first principles we defined in the opening chapter. We can only safely travel from the concrete to the concrete via a looped journey through the abstract. But we also said that it was unnecessary to duplicate research if we could identify where and how it had already been done. We could then lift the results up into the abstract with the same confidence as if we had found them ourselves. Now the absolute-absolute provides us with precisely such an identifiable description and so we can loop it up and away to a more convenient area for experimenting, provided we can define with confidence the abstract path it must take.

Fortunately there is a well developed technique for this leapfrogging of both the absolutes into the new territory without knowing their theoretical relationship, and using the new surroundings to establish it, i.e. dimensional analysis. This is a technique of analysing and relating the quantities by which an object is defined.[14].

This technique can be approached most easily by considering which special category of model automatically retains its reality when uniformally scaled down or up from its normal size. In the football match, this is illustrated by the size of the pitch, the goal or the ball, and almost nothing else. When we are totally concerned with the size of an object and none of its other characteristics a scale model is valid. You modify the absolutes in an absolute ratio, thus transposing reality from one world to another without losing or distorting its identity. The reality–model relationship is valid when you cannot tell by inspection which of the two is the model.

In designing, this technique can be exploited usefully in a number of ways but it has one acute danger. This arises from the fact that often the greatest difficulty about model making is the model maker. He tends to be an idealist: he enjoys the job so much that his dream of perfection may be an accountant's nightmare of cost. Possibly nurtured on model railways where the historical and literal representation of all detail is the goal, he may tend to treat your model in the same reverent spirit. You must discipline him or yourself never to spend a penny merely to make something look pretty. Your money must be solely invested in a replica of reality so simplified that only the essentials remain.

Sometimes (but not often) the exact reproduction of detail is unavoidable, as in the underside and exhaust system of model cars for wind tunnel testing, but be constantly alerted against mere ornamental elaboration.

Probably the best known application of the space model is one that is solely concerned in representing re-scaled linear measurements as in designing the pipework in chemical engineering factories. It is not unusual for the cost of this part of the job only to exceed two million pounds, a sum that may easily increase violently unless the run of each pipe is minutely designed for saving in material and ease of erection. And also pipes mustn't run into each other by mistake. Try this by drawings only, and you start to go mad. A computer will soon be able to do it for you, but not cheaply. Again, if you are concerned in designing the internal contours of a car's front wing to give a stated minimum clearance between it and the tyre over the whole range of possible bumps and steering angles a model will release you from the tedium and possible uncertainty of elaborate drawings.

How late in the erection of a factory can you leave the installation of a large machine and still be able to squeeze it in through other obstacles? Can you remove a car's piston down through the crankcase, will there be enough room for it to pass the crankshaft?

The advantages of models in solving such problems are twofold. The first is that it gives you confidence in your solution; you can see it. The other is that you can equally well see what is not a valid solution, and how much has to be lopped off what, for it to become one.

This 'fitting things in' type of model technique can also show us, in its simplest form, the generalized picture of all types of model making, which is a double-loop into the abstract.

First you have a problem concerned with the design of some concrete object because some essential data is missing. By abstract thinking you isolate this and determine how it may be experimentally established in a full scale rig. If you can define

43

this rig in absolute terms (and, as we have seen, linear dimensions are the simplest form of this), the whole experiment can be realistically scaled up or down via the appropriate abstract theory of dimensions. In short, we can summarize it as concrete reality–abstract thinking– absolute reality–dimensional analysis–absolute model and we can with confidence go to and fro along this double-looped line, and carry our results back to our drawing board.

Now we must turn to the more sophisticated methods of exploiting models. Can we move on from the transposition of linear measurements only to that of total miniature reality? Can we so reproduce a model match that the t.v. viewer can not detect the difference? Yes, by using the theory of dimensional analysis, we sometimes can.[15] Let us first imagine that the dwarfs are half the size of the normal players, but have the same inbuilt expertise at the game. Then, by reducing all the other linear dimensions by a half, making our time go twice as fast and all our objects twice as dense, we can reproduce this indistinguishable Lilliputian reality. The same side will win. Normally research for design using models lies somewhere between the super simple and super sophisticated examples we have just considered. The size of the pitch was a simple item picked from the total picture; the total game was a transposition in depth. What we most often need to do is to select those features out of the total picture that concern our design and, via dimensional analysis, transfer them to a model, subsequently transferring the results of the tests on this model back again via dimensional analysis, to give the full size equivalent in our original picture. But all this is based on the fundamental condition of the relevant realities being definable in absolute terms. Replace any one absolute with a comparable and your reality vanishes in a flash. You must deal in deeds, not words.

This can be illustrated by imagining that you have been asked to design a new pair of goal posts for your local playing field. They must be capable of being removed, as in the summer the ground is devoted to other forms of warfare; and the groundsman must not find re-erecting them too difficult. There are

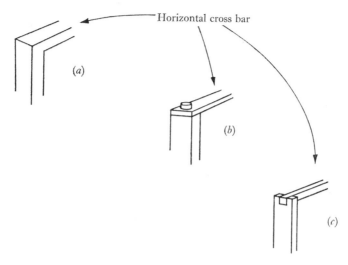

Fig. 7

existing square holes in the ground into which the uprights are to be inserted. In addition, the maximum allowable vertical deflection is specified for the cross bar under a test weight, and a maximum horizontal one when a horizontal test load is applied. You pick on steel as a suitable material and evolve three different designs, as shown in Fig. 7.

(a) Shows a one-piece welded box section structure which is inherently quite rigid. For instance any horizontal deflection in the cross bar is also torsionally resisted by the uprights which cannot turn in their square holes. It is, however, difficult for one man to erect.

(b) Is a three-piece design, where the two uprights can be inserted into the ground in turn and the cross bar then lifted up and its ends dropped down onto the round pins on the top of the uprights. For ease of assembly you must have a fairly large clearance in these holes. It is all easier to erect.

(c) This third design is a modification of (b) intended to overcome the objection that the pins might be bent by rough handling, and so the cross bar is now supported by substantial rectangles fitted into slots in the ends of the uprights.

45

As we will see, these examples are chosen to illustrate our present subject and not to demonstrate profundity of design! Let us also assume that you wish to check your calculated deflections by tests and the use of models would save time and expense.

You would make your three models out of steel, suitably scaled down, and measure the vertical deflections. By the use of dimensional analysis you can then determine what the corresponding deflection would have been on a full size goal. So far so good. Next you apply the appropriate horizontal load (via dimensional analysis again) to each and measure the corresponding deflections, and in so doing you notice that each design is behaving rather differently. (*a*) is, as we have seen, finding extra rigidity from its continuity of construction. By contrast, in (*b*) each member works independently. (*c*) is a compromise, for, at the corners, the ends of the uprights tend to jam in the slots as the deflection increases. In fact these corners start behaving as those in (*b*) but then drift towards (*a*) as they start to seize in the slots. It follows therefore that in all three designs the vertical deflection of the model is a guide to that of the full size goal but for the horizontal deflection only (*a*) and (*b*) are valid. (*c*) has dropped an absolute for you cannot with certainty forecast what is going to happen at the corners. Stored for the summer the ends may become rusty. Taking them down hurriedly to the field for re-erection the groundsman may drop one end of the cross bar with a crash on to concrete and bend it. Trying to straighten it, with the blunt end of an axe, he will make it worse and dent it too. To put it together at all now needs more valiant work with the axe, and the ends become jammed pretty tightly in the slots. But 'pretty tightly' has no absolute definition and even if it had, a shower of rain would wash it away.

A realistic model is ideal, an idealized model is often unrealistic. If you are solely dependent on model testing for confidence in your designs you must drop (*c*) as a possibility; it lacks a vital absolute. It may be the best design but it cannot be verified as such and you cannot risk a guess.

That you may have to restrict your original options to those for which an absolutely valid model can be made is a distinct possibility in design problems of this type. But if you have a realistic model you can exploit all manner of research techniques such as the use of 'influence lines' and so on. This total transposition, via dimensional analysis, is the simplest, but not the only, method of model making. We can use a parallel transformation if this helps.

For instance, we could have done our experiments on a full size goal and if our deflection measurements were likely to be too small to measure easily, we could reduce the cross-sectional dimensions of the cross bar and leave everything else the same. In other words, it would remain full size in one plane, the horizontal, but scaled down in the vertical plane at right angles to it. After testing, dimensional analysis will allow us to re-transpose this new deflection back to an equivalent figure for the unmodified cross bar. But we can go further; we can make a scaled down model of this modified goal, if we want to, and so end up with a total plus a partial transposition, and we can do all this by scaling up and not down if we wish. In fact, if we regarded our total and partial model as the real thing, we can go in reverse and call our original goal a totally partial model of it Or, starting in the middle, we would assume that our modified full size goal was the real one, and on one side we would have a partial transposition, (the old real goal) and on the other a total one (the previous total + partial model). Thus model making can be more than just a straightforward Lilliputian package deal. We can use the dimensional analysis to disect our problems in some detail and run up and down with our scales. But there is one stipulation that must always be preserved. We can only exploit partial transposition if we still conform to our initial structural behaviour. We can reduce the cross-section of goal (b) because it can be isolated from the rest. In design (a) the welded up corners of different basic dimensions would introduce a new design feature and be no longer a replica of reality. More-over, if we reduced the cross-sectional area of the uprights instead of the cross bar, we might cause them to collapse

47

unrealistically early by buckling. Partial transposition depends on our ability to isolate it without ambiguities at the edges.

So far we have assumed that our full scale objects and their models are constructed from the same materials, but this is not essential. We can exploit transformation as well as transposition.

6

TRANSFORMATIONS AND TRANSLATIONS

You are still watching this football match on the television and, naturally, it is a scaled down view that you see. But you don't. If you really had a scaled down version of the match in your sitting room, you would see a number of over-heated little dwarfs running about on the carpet in imminent danger of being eaten up by your dog.

You are not seeing a transposition of the game into your room, but the t.v. has provided you with the game faithfully transferred into another medium, where it re-appears as a small point of light dashing around on a special sheet of glass.

The actions of the players are in fact transformed twice. First into radio waves, then into an optical display; and because we are dealing in absolutes their reality is perfectly preserved. We must next consider how this principle may help in expanding the usefulness of model making.

In the last chapter we assumed that our model goals were constructed of the same material as the full size ones, but this is not mandatory. We can, and often do, transform our model by using another medium of construction; some special material that may help us to save time or money.

For instance, our totally transposed goal models are very stiff and their deflection measurements will require highly precise instruments. Moreover the imposed displacements for influence lines technique are likely to involve considerable forces which again are often inconvenient. If we constructed the model from, say, a suitable plastic, we could make it quicker and test it more easily for the deflections would be far greater and the forces far less. But can we safely do this? Yes, provided we keep three things in mind.

The first is that the stress–strain diagram of our new model material must be made to duplicate that of the original one by only altering by an absolute value its horizontal stress scale.

If this can only be achieved over a limited range of stress and strain, your model must be worked within that range only. If this is not possible, neither is the application of dimensional analysis for we have lost our common pattern of behaviour.

The second condition is that the Poisson ratio of the materials should be the same or similar. The importance of this varies directly with the effect that the Poisson ratio has on the behaviour of the structures involved. Fortunately it can often be ignored. The third condition is that we must construct our model with structural realism. We must stick it together, if necessary, with adhesives that will correspond to a weld, otherwise our absolute precision will collapse, and so, possibly, will the model.

At this stage it is necessary to point out that models are awkward things to devise when matters of self-loading arise. In many a structure a vital problem is to discover how its own weight will affect its own shape. It will certainly deflect, it may possibly collapse. To preserve the realism in a model would mean doubling the density of the material each time you halved its size, and making models out of lead tends to be expensive.

If we can't satisfactorily transform the material can we transform the law of gravity? Yes, put the model in a centrifuge and you can provoke a variable acceleration acting upon it up to two hundred times that due to gravity. Despite the complications of instrumentation it may still be the best way of checking the design of some huge structure. It might be the best way because it was the only way.

At this point we may well look back and review the overall purpose of our model making techniques. Their aim throughout has been to reduce cost while retaining realism. Thus the cost of the centrifuge testing may be colossal, but the alternative of building a full size prototype may be impossible. At this point you may well ask why I am making such a fuss about the cost of the test piece or model; is it not more than likely that the cost of the test apparatus, its instruments and construction, will heavily outweigh that of the specimen sitting in it? And the answer to this introduces a new factor into our consideration of the whole subject.

So far we have always considered our research for design as a means of discovering otherwise uncertain data, or the confirming of otherwise uncertain behaviour. This we have done for simplicity's sake; unfortunately real life is often more complicated. We need data, we need design confirmation, but above all, we need to be able to establish the optimum design for our purpose. Not only something that works, but something that works best, and this implies many experiments. One experiment will tell you nothing about optimization; two will show you only which of the two is the better; you will only find safety in numbers, and the mounting expense of successive models may soon become the major cost. Can we use the principle of transformation to help us in simplifying optimization too? The answer lies in the possibility of electrical transformation. So far we have been exploiting physical transformation. A model of a goal still looks like a goal; a plastic model of it is still recognizable; so far there has always been a family likeness carried across into all our models. Electrical transformation means escaping into a new world altogether such as we have already seen demonstrated in a t.v. picture. There, every physical action on the field is totally transformed into an electrical counterpart in the set. Identifiable reality is reincarnated in another world; nothing is literally the same but everything is analogous. And our first answer to the problem of optimization lies in the total transformation provided by the analogue computer.[16] Our model becomes an electrical model. For any particular physical relationship we substitute an electrical one with the same characteristics. For instance, to take a highly simplified example, the displacement of a structure will depend on its flexibility multiplied by a force, and the electrical counterpart of this might be a voltage, which is the product of a resistance and a current. Thus to construct an analogue computer you must find or devise varying electronic images which reflec the behaviour occurring at successive points in your structure.

Interconnect them, and you will have broken through into a new but analogous world of greatly increased experimental potential.

But we must remember that our overall principle must still be maintained; if we are to travel to a new country it must be by air. We can only safely accomplish the journey by soaring up into the abstract and beaming along established paths of thought.

With a physical model, dimensional analysis was our route, with analogue computers we rely on electronic analysis. Often such computers can beat model making hands down except for one limitation; and it is a very important one.

You can test a model and keep your mind a blank. You are not committed, the model does the work and tells you what happens. You don't even know how the model works it all out; in fact this may be the main purpose of experimenting in this way; the model is likely to be more intelligent than you are.

Unfortunately a computer does not allow you to opt out in this way. It is up to you to find the electrical equivalent of what is going on. Fortunately you can do this a bit at a time; you can send your luggage by air freight in separate little packages, they need not all be squashed up in one case.

In fact, there exists standard electronic packages that provide the analogue equivalent of most things we normally wish to represent, provided we can define it in an absolute way, and this usually means a mathematical one. Put these analogue units together like a game of dominoes and play away. Notice that you have only been responsible for the initial analysing of individual items as such. The total effects of their mutual interaction they will reveal themselves. You can switch your dominoes about or alter their values quite simply and observe how the final result varies, and optimization becomes possible. And this is not all, for electrical transformation has carried with it a transformation of the nature of our exterior loading. No longer do we need to devise testing rigs involving weights or jacks, we simplify everything down to voltages or currents which can be quickly and minutely varied over a wide range. For in-

stance, let us imagine that you are asked to discover the explana
tion and cure of a phenomenon called 'axle-tramp' in motor
vehicles. You will have experienced this yourself (unless you are
very rich and have superior sports cars); and the effect is
unpleasant. Rapid acceleration in the lower gears causes the
back axle to jump up and down; this maddens the passengers,
terrifies the onlookers, and tears rubber off the tyres.

These vibrations are the result of the back axle having to do
two jobs at once. It must both support the car and push it along
and these functions interact on each other. In considering the
suspension system we must therefore include all the movements
that are likely to occur, including the fore-and-aft displacement
of the front-end of the propeller shaft and the corresponding
one of the flexibly mounted engine gearbox unit, as well as the
five degrees of freedom of the axle. In the transmission system
we have to include the flexibility of the tyres, the torsional
flexibility of the half shafts, propeller shaft gearbox and clutch,
the driven plate on the clutch and the effect of the engine
flywheel.

To define these, seventy-eight variables are necessary. The
whole system is then acted upon by five different external force
functions all of which must be varied over a wide range. Model
making is impossible; it would take several lifetimes to explore
fully all the combinations of models and conditions. To use an
analogue computer you need only work out that there are eleven
degrees of freedom, put each in a mathematical form and pick
out or build up their electrical counterparts. You will end up
with about seventy interconnected standard analogue units, and
you can then play away to your heart's content, your axle-tramp
appearing in graphical form. You will be able to find what is the
most important factor in preventing it, i.e. the longitudinal axle
mounting stiffness, and design your suspension to optimize
this.

If you think that my selection of an illustration with seventy-
eight variables is an extreme one you may well be right. It is, by
contrast to many, extremely simple. My own experience includes a problem containing 3,000 variables but others will

doubtless know of much more complicated problems than this being solved quickly by computers.

Meanwhile, the football match has ended and the t.v. screen shows the final score in big superimposed figures; someone has won by two goals to one. You can, in fact, discover who won in two quite different ways. You can watch the real game and see what happens. Alternatively you can learn the same fact translated into digital form; into the two numerals. The match, and the transformed reproduction of it that you were watching, were real and absolute. You saw it. If you only observe the score you remain in the abstract. You never see the game; you just see the result. Numbers are abstract; you can see two goal posts but can never see 'two' by itself. The winning shot was real, but the final score was abstract. When we fly up into the digital world we have to stop up there; a down to earth digit is a 'self-contradiction'. Now this principle of translating reality into the digital abstract has a great potential for the designer. Not only can a digital computer deal with digits at lightning speed, but if you translate reality into these terms you have an abstract world at your finger tips, and thus you have found a sphere directly compatible and often receptive of your own abstract thinking, i.e. your mathematics and your memory.[17] As we shall see later, the designer and his computer can converse with each other. And this brings us to another distinction between analogue and digital computers as a design aid. A digital computer has no direct dealing with reality; it just sits there and does arithmetic. You are its only contact with the world outside. You must tell it how to do its job. It is splendid at doing sums but hopeless at doing mathematics; you must do all that for it. Almost all mathematics can be reduced to digits if you know how, and unless you know how a digital computer is useless except for putting cups of coffee on. It can learn anything but originate nothing. In contrast an analogue computer has, to a limited extent, a mind of its own; or rather, a number of little minds passing on ideas, which may come up with something that never occurred to you, and present it in a form that you can more easily appreciate. For instance, I was once faced with the problem of trying to discover if a series of results had some hidden pattern or if they were totally random. I found the solution by using an analogue computer, in which I arranged that each numerical result was represented by its own note of music. All I had to do was to listen and see if the computer played a tune. It did.

7

THE DICTATORSHIP OF TIME

We camp with Newton on a special ledge sliced out of the Matterhorn of reality and our experience of this rugged and unpredictable world around us can be interpreted by this fact. As we have seen, we are specially equipped by our senses to live on this unique level. But can we define the basis from which this uniqueness springs?

The answer lies, I suggest, in the fact that our scale of experience lies within that no man's land of physics where relativity and quantum physics intermingle. And the characteristic of this no man's land, where all men live, is that it is under the dictatorship of time. In other spheres time is often a partner, even a junior partner, but with us he is the boss. We cannot opt out from his edicts, our most homely phrases reflect his dictatorship. We talk of 'half-time' and 'overtime' and alarm clocks and calendars are unlikely to become extinct. None of us can, as yet, escape from that rugged ruler, the ancient with the scythe.

It is the time element that makes life so complicated, and in particular, often makes our research complicated too. We cannot avoid its domination ourselves, but can we filter it out from our experiments? Yes, we sometimes can and, when we can, we must, for the savings are notable.

There are two possible escape routes from time's imprisoning complexities; we must organize our variables so that they either have no relationship with time at all or else have an identical relationship with it. The reason for this can be explained by the basic concrete–abstract–concrete principle defined in our opening chapter. For variables to continue to share a common area, they must either sit tight and so avoid travelling through the abstract at all, or else they must hold hands and journey together. We must now insert these principles into our overall concept of research and design. As we have seen in the preceding chapters, we can do full scale research without having any ideas in advance about what is going to happen. We build up an experimental rig, actuate it, and watch what happens with varying degrees of apprehension. Neither we nor our ideas are necessarily involved in the experiment; we solely spectate. The next step was to simplify this free-for-all set-up by disciplining or eliminating variables to give a more focussed result. Then we turned to the economies in time and cost by exploiting models via transposition and transformation, and the examples we used had one common characteristic. They were chosen because time had no decisive hand in moulding them. We started with an example of simple measurement; there was no time element in the size of a football field. It just sits there all the time. Then we considered the design of goal posts when loaded in a certain way. Although 'when' is the usual way of expressing this, it would be more technically accurate if we said 'while' they were loaded. There is a time element in both the load and the deflection, but, as it is the same for each, we can forget it. We can totally transpose such experiments up or down in scale, and pretend, with a free conscience, that time is non-existent.

Such examples are useful for illustrating as simply as possible various experimental techniques, but they are over simplified as a mirror of our often time-orientated problems.

This can be seen by reverting to our match, where the goal keeper has let through the last eleven successive shots at goal. His enraged team mates are now knocking his head against a goal post. If they do this with sufficient enthusiasm and rapidity the goal will start vibrating at some natural resonant frequency. The simple minded world of fixed forces and constant deflections is suddenly invaded by time's unruly children named velocity, acceleration and momentum, and their disreputable friends who climb over everything. If you have thoughtfully attached suitable instruments to the goal in advance you can record the resulting chaos; you can record it, but you cannot safely transpose it. Bribe a number of dwarfs on a miniature field to do the same thing and chaos will again arise, but it will be multiple chaos for it will be unrelatable to the original one. This is because the new variables are not all related to time in the same way. Reduce the scale, and the forces of gravity and the composition of the air are unaffected and retain their old relationship to time. The dynamic characteristics of goal posts do not, and what happens to the damping effect of the posts in their holes is anyone's guess. I remember how startled I was the first time I found that, in some cases, damping influenced resonant frequency. And there is no foolproof solution, in principle, to this problem. You can't dictate to time; you have to obey him, and major problems arise from his frustrating embargoes.

If you are selling a motorcar you can take out a prospective customer for a ride in a demonstration model first. Then, if he likes, he can buy an identical model. If you are selling a power station you usually cannot show it off first. Power stations are sold to a promised future specification, and it would be of enormous help in selling it if you could confirm the design by testing it on a small scale working model first. An increase in efficiency of 0·1 per cent in a power station is worth more than £100,000 in its capital cost. But you cannot establish total efficiency in advance in a scale model; your complex variables have no common conformity to time, and no one has yet found a way to by-pass the difficulty; a fortune awaits anyone who can.

If time's egg-whisking hand has made the total transposition in models an impossibility in many realms, often including those where it would help most, can we profitably follow the pattern of the previous chapter and attempt partial transposition? Yes, to a certain extent. The extent depends on the degree to which we can find a suitable relationship amongst our variables. The clue to finding one depends on the possibility of exploiting for our own advantage some of time's bedevilling complications.

When we were limiting our attention to static loads and measuring deflections we found that in making a Lilliputian model everything became scaled down; not scaled down necessarily in the same ratio but at least nothing became larger than life size. This is not so when time starts bossing us about, for anything can then happen. If you make a model of a goal; the size goes down but the resonant frequency goes up. We do not necessarily know which way things are going to jump. Nonetheless, if we can isolate some of them we may be able to manoeuvre them into a helpful pattern.

Let us imagine an island of reality, A_1, which we wish to scale down into a model of size A_2. Using our basic concept of the concrete–abstract–concrete loop we use dimensional analysis to provide the connection.

Next suppose that in some other area of reality we have another island, B_1, which we wish to scale down similarly. It travels through its own abstract path in a corresponding way and arrives at B_2. Now we will assume, that, by a coincidence, the abstract path followed by B_1 is a similar shape to that which A_1 has traversed. This means that the ratio of their original sizes, A_1/B_1, remains unaltered by the scaling and so A_2/B_2 has this same value. Two things follow from this. The first is that if we are told that in another model, A_3 has a certain size, we can immediately fix the value of B_3, for A_3/B_3 will remain the same as A_1/B_1, and we can calculate this fraction without having to journey through the abstract or even know the shape of the route. It is still the ratio of their life size dimensions.

The second point is that this original value of A_1/B_1, and any other scaled combination of them, always remains at a constant figure, and this figure has no dimensions. The dimensions have all cancelled out. If sharing a common abstract, it follows that the quotient of any scaled A and B will be a number only. We can often compress all this into one sentence, by saying that if two realities have a similar relationship with time (or its by-products) they will have a constant relationship with each other, independent of scaling.

Earlier we discussed how real complications can be made to cancel out, here we exploit the same principle with abstract ones.

Now if, by another coincidence, we find that another variable C is always absolutely dependent on the ratio of A and B, C will always have the same number at any scale of modelling.

By much the greatest, the amount of money spent on research

for design lies in the realm of fluids i.e.: aircraft and ships, and it would be a great deal more had not this technique of dimensionless numbers been exploited. In most of them 'A' and 'B' are some type of force and inertial effects (the Mach number is an exception) and they allow us to isolate out from time's hubbub a few clear statements. We plan a series of models around these dimensionless numbers, or combinations of them, and put the results together to establish a clear statement of the total reality, or as much of it as we can.

In the previous chapter we likened the step-by-step progress of an analogue computer to a game of dominoes where the pieces are placed end to end.

In fluid mechanics research we do not put them end to end but try to fill a vacant area with them as best we can, like a jigsaw puzzle. Each separate experiment may give us a piece, but you cannot guarantee in advance that you will have all that you need, nor that some of them will not turn out to be double blanks.

Successful research for design may well turn out to be a combination of your tests, and other people's mathematics, all spaced out by some lucky guesses.

Are you following all this?

'No, I'm miles behind totally lost beneath a heap of absolutes, abstracts, dominoes, and so on, all tied up in abstract loops.'

I'm sorry about that, but perhaps you are not really concentrating.

'Possibly I'm not, I've too much on my mind. I must do some experiments urgently and I'm skimming through this book to find some example that I can copy.'

Well, I hope you don't succeed. It would almost certainly be a highly unsuccessful success.

'Are you trying to be helpful or just confusing?'

I'm trying to stop you confusing yourself and wasting your research allocation. This book is about general principles of research illustrated by particular examples. Don't try to stand it on its head. To find an exact duplicate of the experiments you need to do is statistically very unlikely indeed. You may find something that is 'nearly' the same, but 'nearly' and 'the same' may be a long way from each other and only an appreciation of general principles will enable you to judge just how far. In this subject, if you try to get rich quick, you get poor even quicker. Research is always, to a certain extent, a gamble and you must try to reduce the odds as much as you can by disciplined thought. I have been told that in fundamental scientific research, efficiency is of the order of five per cent on the average; if the right thing was always done first time the efficiency would be 100 per cent, but you should do much better than the average.

'Why pick on me?'

Because you are doing a specialized form of research, compared with other realms your objective is more limited and defined. You want to exploit an unknown area in a known country; the fundamental scientists go out to exploit the totally unknown. You know what you don't know; they don't know anything. But with this advantage of a known objective you have, almost certainly, the liability of a limited time or limited money, or, most probably, both. In some instances time is so important that firms are prepared to spend large sums of money to reduce it, for the design of large projects. Recently I was allocated a million pounds to spend in twelve months for such a purpose but this must be chickenfeed compared with some people.

'It is exactly the opposite with me, I've precious little money but everyone wants the results yesterday and if the design is a flop I shall be in the dog-house.'

May I ask what is this design problem that you keep muttering about?

'Have you ever heard of a Roots blower?'[18]

Yes, they were invented before even I was and have been widely used in marine, automobile and aircraft design as well as process engineering. What's your trouble?

'I must specify the design of one and be certain in advance of the performance it will have when built and how this be-

61

haviour might be modified by altering clearances, sizes and so on if the sales people wanted it.'

Show me the basic design and tell me how it works.

'It's all quite simple; look at Fig. 8. The two rotors are geared together so that there is a small clearance maintained between them, they snatch in air at one end and fling it out at the other like all constant displacement pumps. I don't know where to start on the problem of designing one.'

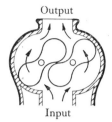

Fig. 8

Start by understanding how it works.

'I've already told you that.'

No you haven't, you must work out exactly how it works; your vague description is quite unrealistic and will mislead both of us if we are not careful. If you don't start with reality you are unlikely to end up with it. Forget about all that snatching and flinging and tell me what really happens. You had better use a diagrammatic drawing.

'I hope Fig. 9 (1) is what you want. The air enters the inlet port and fills up the area "A" between the rotors.'

This air is at normal atmospheric pressure I assume?

'Yes: next the rotors displace themselves to position (2), where the air is trapped in area "B".'

Still at atmospheric pressure?

'Yes, I suppose so.'

Don't suppose things, Yes or No?

'Yes, provided we ignore possible leakages backwards from the output side through the various rotor clearances.'

That's better, now what happens.

'The rotors reach position (3) and a small gap opens up into the delivery pipe and out the air goes.'

Now you are getting into a muddle again. You said the air in 'B' was at atmospheric pressure, while the air in the output area must be at a higher pressure than this. If it wasn't it wouldn't be a pump. You tell me what happens next.

'Well, I suppose that immediately the gap opens, the air in the output side, being at a higher pressure, will flow into the area B.'

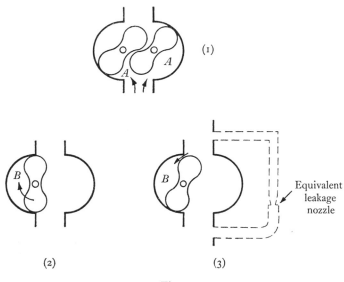

Fig. 9

Will it or won't it, don't keep saying 'suppose'.

'It will.'

So, at this point, the air, instead of being 'flung out' actually moves in the opposite direction and flows into the space between the rotors and sweeps away your remark about it being a 'positive displacement pump'.

What happens to the air next?

'I supp—I mean it will travel across, as shown by the arrow, in the form of a pressure wave, until it hits the tip of the second lobe. In practice the lobe speed will be an appreciable proportion of the speed of sound and so the returning pressure wave, reflected back from the lobes, will burst through the enlarged gap into the output space where it will eventually expand and, in doing so, will send back a depression wave into the pump and so on.'

Yes, there is obviously never a dull moment, but how will this impinge on your design ideas?

'Well, I thought a moment ago that the gears locating the

63

rotors had a straightforward power transmission job to do but this obviously is not so now.

'The smash and grab raid of pressure and depression waves are going to react against the gear teeth, distort the rotors slightly and set up torsional vibrations in the shafts. Everything will have to be stronger and better designed than I anticipated. You don't seem to be making things any simpler for me.'

No, I am forcing you to be realistic which often means facing up to complications that, subconsciously perhaps, you previously were apprehensive of recognizing. It is a complicated situation, but not nearly such a complicated one as a pump that won't work, or disintegrates.

And this brings us to the overriding difficulty of designing in this realm, where time's delinquent offspring seem to relish disturbing the peace. It is that everything is so muddling and involved that you don't know how or where to start. Now if you had not been merely skimming through this book looking for illustrations you would have noticed that I have already tried to advise you what to do in this type of situation where variables proliferate and scale factors abound.

'Actually I do seem to remember something about that. Was it that I could only fill in the picture by a mixture of lucky guesses, mathematics and tests?'

Yes, and this gives you a marvellous and appropriate opportunity to use your favourite word.

'What's that?'

Suppose.

'Well I suppose it is, but you are making me self-conscious about it.'

Sorry about that, but what I wanted to do was to make you conscious of it. The ability to suppose things is a vital part of a researcher's equipment and all the more useful if it is consciously exploited. Paradoxically this can be done in two apparently opposing ways. The first is to suppose that matters are more involved than you imagine that they are, and then you seek to find out if your supposition is correct. The illustration of the seesaw, Fig. 1 is an example of this. It is easy to form a mental picture of a stationary seesaw; in fact that is what most people would probably do. But if you say to yourself 'Suppose I am wrong, suppose something else happens, suppose it behaves like a moving seesaw.' You will then, having broken out of your simple minded attitude, use a prototype, a model or a computer to discover if your guess is correct. Such suppositions are an insurance against the unexpected.

The other alternative is to assume the opposite: i.e. matters are simpler than they appear. This is an invaluable, and often the only, means of achieving some degree of order in a hopelessly involved situation, and this Roots blower is a typical example. You will have to list about half-a-dozen guesses, preferably good ones, if you are to save time and cost. If you have unlimited quantities of each, you could dispense with suppositions. As you haven't, guess away.

'Can you give me a start?'

Yes: I think the first thing you must suppose is that air behaves like a 'perfect' gas. This type of assumption is probably the commonest one of all in research; it means that a material behaves 'perfectly' or 'exactly like' the standard mathematical picture of it: In this case the assumption is extremely useful as you can immediately pick up and apply a host of ready-made mathematical tools that other people have prepared for you. The less thinking you have to do, the better for you and everyone else. Looking at it the other way round, you could say that if air cannot be assumed a 'perfect' gas in this sense, no exact mathematics can be done on it, and so model making cannot be usefully exploited and you are forced back on to high costs and long times. You guess the next one.

'I suppose for similar reasons, we must assume that processes are adiabatic.'

Yes. You are getting the idea, and have filled up two of the blank spaces in the total picture. Try and find some more by looking more closely at how the pump works.

'As the pump rotates the pressure must fluctuate in the receiver it is pumping into, but this depends on its size. Could we eliminate this variable by assuming it is of infinite size?'

Yes, you will have to do that; now what about all these air leaks that must arise; not only because of the space between the rotors, which would seize if they touched, but also down their sides and over their tips?

'Can we lump all these leaks together and assume that they are equivalent to a small return pipe (shown dotted in Fig. 9 (3)) with a convergent nozzle, whose isentropic performance is mathematically well known?'

Yes, you seem to be getting the idea now, and it will be worth summarizing it in a general way which I hope will go on helping you when Roots blowers are a thing of the past.

The science of design must almost always be, in practice, the science of disciplining variables. As we have seen, this problem, wearing many different disguises keeps popping up and playing the villain, never more so than when the stage is uncomfortably full already.

If we cannot buy him off with money we must take the sting out of his performance by cunning and the method should be this.

Faced initially with a complicated situation, climb down into it to a greater depth to make sure whether it may not really be even more complicated still. Then having found the realistic level for the stage, try to substitute characters you can be sure of for any unruly ones; this can be done in two ways. First try to replace an unknown variable with an unvarying variable, i.e. a constant, thereby turning a character into a stage prop. Secondly, try to replace unpredictable characters with actors you already know well and who always put on the same act, a stereotype performance of mathematical predictability. Naturally, all these manoeuvrings will modify the plot somewhat and you are forced to gamble that the general effect will not be misleading.

Now having done your best in this direction, and incidentally reduced the variables in the Roots blower problem from over fifty to twenty, you must now attempt to make sense out of the plot. How are the variables interacting on each other and scale on all of them? As we have already discussed, in this context there are no certainties except mathematical ones.

66

Searching for some simplifying gambit, you must try to find some repeated pattern, however indistinct, if you can. A repeated pattern can sometimes be manoeuvred into cancelling itself out. A grandfather may have two grandchildren of different ages with different parents, their ages and the ratios of their ages will vary with time, but define them as cousins and you have a verbal relationship unaffected by time at all. Have a look at the diagram (Fig. 9) again and see if you can spot any grandchildren of old Father Time.

'It's rather difficult but the air pumped through seems to be dependent on similar general conditions as the air leaking back through our imaginary pipe.'

Well it's only a shot in the dark but see if you can find a direct non-dimensional relationship between them, by juggling about their mathematical backgrounds.

'It takes a bit of working out but these two are related by a non-dimensional number which in turn can be related to volumetric efficiency, blower speed and pressure ratio. If you put all four of them on to the same graph the effect of varying any of them can be simply seen. It's rather fun.'

Yes but beware of the danger it carries with it, the often subconscious feeling that, if a phenomenon can be expressed in an elegant mathematical way, the mathematics is likely to be realistic. Watch out for wishful thinking. The point is not whether the equations are fun but whether they are true. Remember that reality lies on this side of the footlights. We have modified the plot into a triangular drama played by three characters and a stooge with a walking on part and entertaining it may well be, but you must now find out if they are only playacting.

'How do I do that?'

By exploiting the third factor I suggested earlier (page 60) i.e. by realistic experiments. Take a Roots blower, measure its performance in the terms of your graph and so put in a real value, then its performance at other loads and speeds can be absolutely predicted, and scale effects too. Although in theory one test is enough, you had better do several at different speeds to check the mathematical consistency of the concrete points.

Looking back you can see that you were initially faced with a problem where their were about fifty variables and a possibility of having to do thousands of tests to sort them out. Now you are down to four variables and a few tests, but the important point is the general pattern of thinking that achieved this for you. It can be applied over vast areas of research for design.

8

LOOKING, SEEING AND BELIEVING

'Science is concerned,' says the Oxford Dictionary,' with observed facts.' But there is more in this than meets the eye. Seeing is said to be believing but in the science of design beware of believing all that you see.

'That sounds most impressive, but I haven't a clue what you are talking about.'

Look at Fig. 10 and tell me what you see.

'A circle with a point in the middle.'

Exactly; now tell what it represents if you imagine it to be a three-dimensional object.

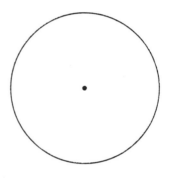

Fig. 10

'Well, it could be the top view of a dunce's cap, a disc with a hole in it, a coin in the bottom of a well, a distant football seen through a porthole; dozens of different things in fact.'

In other words, looking at Fig. 10 in two dimensions you can define what it is; in three dimensions only what it *might* be.

The next thing I want you to do is to say to yourself 'I am going to see a disc with a hole in it' (or any of the other possibilities), and then look again at the point. You will then find that you will see it as the thing you were anticipating. Now from all this you can deduce three fundamental facts about observation:

(1) You can define exactly a two-dimensional figure.

(2) You cannot define exactly a three-dimensional figure.

(3) With (2) what you will see will be dictated by what you want or expect to see. In other words you are believing yourself, not believing the object. From this follows:

(4) A two-dimensional figure tells you nothing about itself in three dimensions.

'Is this supposed to be helpful?'

Yes, but my point at the moment is that all this is inevitable, the structure of your eye makes this so. Light is reflected from an object and falls on the retina of your eye. The image it forms is a two-dimensional one, so the only information your eye can receive or pass on unambiguously must be expressed two-dimensionally. With the other eye supporting it stereoscopically a three-dimensional picture can be built up but only after the mind has evolved a computer program for doing this. A baby can look but not see, for seeing is interpreted looking. By feeling objects he can develop his interactive program and soon begins to see them too. This ability to promote a computer-like program when required is a gift that we never grow out of. Put on a pair of spectacles that makes everything you see look upside down and in three weeks you will be seeing them the right way up. A new program has been hurriedly written. Take your spectacles off again and it takes another three weeks to get back to the old program. All this goes on subconsciously and pragmatically. Often, when a child, our subconscious mind is much cleverer in its achievements than the conscious mind is. We soon discovered that toy bricks were not all the same weight, and we had to hold tighter and brace our arm muscles more for the heavier ones. That weight was volume multiplied by density was an abstract idea far beyond our conscious thinking, but our subconscious immediately got down to work on the problem. It soon related the differing densities of wood, metal or liquid in a cup, to their appearances. Even now if we pick up a large block made of balsa wood we feel a sudden shock; it is so much lighter than our program says normal wood should be, and a quick correction is necessary. Even more difficult, we learn to assess volume from looking at size, and again we get that sudden shock if the object is unexpectedly hollow.

But we must remember that, clever as all this is, it is a blind cleverness. It is built up from looking, not seeing, it guesses to find a guess that works. And if it can't check a thing it may make stupid mistakes. A child learns to fan himself with a piece of cardboard or a small leafy branch from a tree. G. K. Chesterton, when young, thought the wind was caused by the trees fanning the air about. Looking, interpreted into seeing, i.e. two dimensions programmed into three, is potentially highly useful or highly misleading, and can sometimes be highly entertaining as well. Conjurers rely on persuading us to put in the wrong program. In Hitchcock's film 'North by Northwest' we see a plane fly full tilt into a huge lorry. This is what we see, what we are looking at is sometimes quite different. The film was made by organizing a low flying aircraft to skim over the fields straight at the lorry, then rising enough at the last moment to miss it. From the filmed sequence the last minute lift of the plane was cut out and it was joined directly on to a sequence showing a stationary plane smashed and landed on a wrecked lorry. We 'see' the crash, for we have no alternative program marked 'Hitchcock's at it again' to slip in. It is the absence of a suitable program and the hurried insertion of the nearest that can be found that is often the most misleading of all. If you are shown an object and then asked to picture what it would look like if made a quarter of its original size you would almost certainly get it wrong. You would picture it too large. As such linear scaling has no normal utilitarian use, we have not programmed it. When asked to do it our subconscious searches wildly around and picks on the nearest it can find, which is the old volume judging one we use for weight lifting, and so we 'see' in our imagination an object which is not a quarter of the size but a quarter of the volume.

Accurate observation consists in the accurate exploitation of accurate mental programming. But this is not as easy as it sounds. It is rather odd, but a fact, that we are poor judges of the efficiencies of these environmentally-based faculties. Sometimes we are better than we think we are, more often we are worse.

For instance, when we listen to someone talking to us we watch their face. We therefore develop a program where our eyes help our ears, for if there is a large amount of extraneous

71

noise going on, we can help select the sounds we need to hear by watching a person's lips move, and we are better at it than we generally realize.

Imagine you have taken your best girl to a party, where there is the usual din going on, everyone talking hard against a background of howling music. She has wandered over to the other side of the floor and is conferring earnestly to a man in the corner. You would desperately like to know what she is saying, and even more, what he is. But politeness prevents you plunging across the floor after them. What should you do?

'I've no idea but I'm all ears, this is what I call vital research.'

You'll never learn anything if you are 'all ears'; the trick is to watch intently the movement of the lips of the person you want to hear, and this will enable your ears to focus on the sounds associated with them and you will discover what those two are plotting.

'Thank you very much, it is the most useful thing you have told me so far.'

By contrast, we more often over estimate our abilities; an example of this was recently seen in a plane crash. This happened at an air display when, tragically, the aircraft began to disintegrate in the air and then crashed.

To help investigate the cause, people were asked to write in and describe what they saw. Over a hundred people did so. Then it was discovered a news film had pictured the crash and all that led up to it. This showed that none of those who wrote in had observed accurately the sequence of events. Of course they would have done better if they had been told in advance what was to be expected of them; thus they would have inserted the appropriate 'Memorize this' program in readiness.

This brings us to the fact that switching programs cannot usually be done without a finite delay. An illustration of this can well be that of the circle and the point we have already looked at. The first three-dimensional shape it suggested to you was probably subconsciously selected by the nearest program you had to it, or the last one you were using. 'It is a dunce's cap' you said to yourself while looking at it. 'Of course, we can see it is' your eyes replied. And they will go on seeing it that way. If someone now says to you that it is actually a coin at the bottom of a well you may have real difficulty in switching your interpretation across to this immediately. You may well have to look away for a moment or even to another drawing; then you will be able to say 'It is a well with a coin in it' and your eyes will reply 'of course it is'.

Now in observation this inability to change programs quickly is often awkward. I remember an occasion when my research team and I were concentrating on the minute movements of four dials and suddenly something totally unexpected happened and all the indicating hands began jumping about in an alarming fashion. Afterwards none of us could recall how it all started. We did it again and, alerted in advance, had no difficulty.

Finally there is something that we are all bad at and probably know it. Unless compensated by special training our minds are hopeless over units, especially the unit of time.

'What, that old fellow being a nuisance again?'

Yes, it is surprising how he keeps butting in and making problems. Some years ago a steam turbine generator disintegrated in a spectacular manner causing extensive damage. In the enquiry it was established, by examining the various bits and pieces that were left, that the most likely cause was that the speed control had failed and the turbine rotor had been burst by centrifugal force. In theory, the time taken to reach the bursting speed from the normal running speed, if unrestrained, was thirteen seconds.

In the event, the fact that something unusual was happening was confirmed by the expert staff who immediately recognized the sudden change in the note of the noise from the turbines. 'How long' they were asked 'was it from then to the explosion?' Each gave the time in seconds, and their figures ranged from eight to thirteen. The answer was vital; a time appreciably less than thirteen seconds meant that some other factor apart from centrifugal force had to be found, but no one could agree about the precise time.[19]

'I am not at all clear how all this fits in with the previous chapters; you have never mentioned observation before.'

I am glad to see that you have read a few pages without skipping; I thought that dragging in your girl friend would do the trick.

'I suppose so.'

With regard to its relevance with the rest, you will be able, possibly, to understand this better if I ask you to imagine that you are observing an elephant. The elephant and the observation of it are two different things. If you go your observation disappears into thin air but the elephant remains there, hale and hearty. The elephant is in a zoo, the observation is in your mind. The first is concrete, the second abstract. Now my basic theme, from which everything is deduced, is the concrete–abstract–concrete loop, and a vital link in this is the frontier between the concrete and abstract and the smoothness of the traffic flow across it. 'Observation' is the customs post; at this precise point we pass from the concrete to the abstract. The scope, speed and efficiency of the whole is largely dependent on what happens here. I have tried to point out that dependence on our natural abilities alone is going to cause a snarl up of traffic with indefinite delays and misunderstandings. Our powers of observation need some outside assistance, and this is most usually done by the use of instruments. This is the next topic we must tackle.

9

THE CONSORTIUM

In regard to instruments, natural scientists and mechanical scientists have a common interest but a contrasting attitude.

To the natural scientist bigger, better and often more expensive instruments are stepping stones to a triumph, while to the engineer they are more likely to be a short cut to bankruptcy. A natural scientist is often chosen for his ability in evolving instruments as much as for that of being able to interpret what they tell him.

An engineer, if he is wise, will avoid inventing new instruments like the plague (unless that is his actual job) and for three very good reasons.

The first is that he will, almost certainly, be bad at doing it. It is a highly skilled specialized profession where amateurs are likely to make fools of themselves.

Secondly, even if he designs and makes an original one he cannot be sure of its limits of accuracy without a long, probably very long, series of tests. It will be an expensive red-herring.

Thirdly, he must, for the sake both of his own sanity and his sponsors' money pick other people's brains, experience, and production lines as much as possible. He cannot be a one-man band and it is stupid to try.

Don't make instruments, buy them. If you cannot find exactly what you want, modify your research set-up until you can. Only at the very very last resort make your own, and even then do it by a combination or modification of existing ones. You have enough trouble of your own without taking on any more.

This brings us to the question of how instruments should be chosen and for this we must rely a good deal on what we discussed in the preceding chapter.

The value of an instrument, in practice, depends on its ability to latch on to an existing and informative mental program in the observer. It is the clarity of the total resulting programming that counts and one on which the efficiency and reliability of the take-off from the concrete to the abstract so depends.

At first sight you might think that the more accurate the instrument the more effective the consortium of researcher and instrument is likely to be, but this is not necessarily true.

Take, for example, the first demonstration that is usually shown to beginners as an introduction to research into materials.

A length of mild steel is placed in a tensile testing machine and we are told to observe what happens and, in particular, to notice how, at one point, the steel stretches quite rapidly without any great increase in tensile force. This is called the yield point of the steel and the stress at which it occurs is most important.

We will assume that the amount of elongation in the steel, measured from its initial state, is simultaneously shown by four different instruments. The first of these is a graphical one, where the elongation rotates a drum with graph paper wrapped horizontally round it, and an inked pointer, moving vertically, draws in the relationship between elongation and the applied force.

A second arrangement measures the same two factors by showing their values on two dials, where the indicating pointers move over an angular scale.

The third set-up is the same as the second but the elongation is now measured much more accurately, with an extensometer in which the indicator is continuously revolving, each rotation representing a small increase in length or force.

Fourthly we have both the elongation and the force indicated by digits, the values being represented like the mileage indicator on a car.

The test now begins and we are told to watch for the point at which the steel begins to yield.

The first instrument, the graphical one, shows this as a sudden plateau cut into the smoothly climbing curve drawn by the pen.

In the second it is indicated by the momentary hesitation or stopping of the force indicator while the elongation one moves more rapidly.

In the third one, with its sensitive measuring, we can only recognize the yield point by spotting the force at which the extensometer stops behaving like a slow electric fan and turns into a fast one. With the digital display, we must keep our eye on the two increasing values and spot the point at which their arithmetical ratio begins markedly to alter.

Of the four types, the first instrument, the graphical one, naturally gives the most unmistakable information about the yield point. Is there anything else of interest you can learn from all this?

'Is it that the more accurate the instruments are, the less accurate is the total effect?'

Yes, you are paying attention for once, keep it up and try to answer something else.

Suppose we had done the same experiment with the same four instrumental set-ups; but we were intent on discovering the modulus of elasticity of the steel specimen. This the ratio of the elongation to the force at stresses lower than the yield point, where the steel is much stiffer.

Which instrument would have done best for this?

'Well, the graphical one would be hopeless for we would have to measure the angle of a nearly vertical line. The others would be better in turn but easily the best would be the fourth; it would have the figures worked out for you already. You would just have to do the arithmetic.'

Yes, in other words the order of instrumental effectiveness has been exactly reversed, and a study of why this should be so will introduce us to the general technique for choosing the type of instrument we require.

The first type, that of a recorded graph, has the advantage that a sudden change in shape such as the ledge occurring in an otherwise smooth curve at the yield point, is readily apparent. This is because it exploits an existing mental program that we all possess. From the days of our toys we have been able to test by touch the smoothness of an outline and so build up an optical interpretation and recognition of it.

There are, in fact, three different types of environmentally

77

based gifts for appreciating changes in form: at the first of which, the one we have just defined, we are all good; the second depends very much on the individual; and the third is one we are all bad at doing. The second type is concerned with recognizing a step on a surface in three dimensions; the first type was concerned with only two. As we have already suggested, the efficiency of our built-in programs is conditioned to a marked extent by our childhood surroundings. An Indian child, growing up in Britain will play with toys of determinable and fixed shapes (i.e. bricks etc.) and, if not carefully watched, will read comic strips in children's magazines.

A British child, brought up in India, will play with flowers, mud and water (which must be great fun), and fortunately never see a comic strip. Moreover Western art has embodied perspective for centuries, while in Indian art forms it is rare.

Thus the Indian child, in the British surroundings we have pictured, will have an initial advantage in picturing three-dimensional shapes. Subsequent practice in visualization can even things out but, even so, individuals vary widely. It is a matter of luck, or hard luck.

Thirdly, there is the type of observation that we are all bad at making; that is the creative visualization of surfaces. If a surface exists we can estimate its smoothness; our instinct for self preservation has compelled us to learn this. Without it we would always fall down stairs and never be able to carve the Sunday joint.

Moreover we absorb this gift automatically from nature direct, as we look at trees and see how the branches join the trunk in various degrees of blending. We build up an artistic appreciation that finds fulfilment in many ways such as sculpture or the highland mountains.

But this does not equip us for imagining what a surface should be like when we can't see it. Try and visualize the exact shape of a junction of three offset round pipes and you are lost. Try it with only two and you may still be lost. We are good at assessing the smoothness of surfaces but hopeless at the mathematics of them. Fortunately computers are excellent at doing this bit, and can thus provide the basis of a useful and sometimes indispensible partnership. Program the computer to do the donkey work on surfaces to a level at which it will latch into our mental programs on smoothness and we can save two years in designing a plane.

Our next type of instrument was that of a pointer moving over an arc. It conveys its information with little conscious effort as it enacts a familiar phenomenon. We are unconsciously wary of trees that sway in the wind and things that are beginning to topple over and thus our reactions are quick in this familiar context.

In theory, I suppose, an indicating pointer with its bearing at the bottom may be fractionally better because it is fractionally more natural than the pendulum-aping top bearing alternative. Isn't it slightly quicker to read the time when it says five minutes to one as compared with reading it at thirty-five minutes past six? Next we come to the rotating-type dials. Although they are hopeless to read while moving, they can be as precise as you can reasonably ask when stationary. That we find them uncommunicative when behaving like electric fans can also be explained by their lack of a natural counterpart. Nothing goes observably round and round in nature, except the world itself, and this we never notice as the sun ducks out of sight and so conceals the fact. Man had to invent a wheel because nature had not left any lying about. If we feel that attempting to observe a revolving dial is unnatural, it is because it is unnatural. Thus the short answer to instrument selection is to choose the most 'natural' one. Thus pattern recognition which nature teaches us in streams, flowers, and a multitude of glories is usefully exploitable, as in photoelasticity, interferometry and kindred techniques.

Lastly we have the instrument that registers the elongation in digital form; it is almost certainly electronic in construction, a six-figure readout is quite usual and, in theory, the sky, or rather the wavelength of light is the limit. Large numbers are almost as readily available as small and this introduces us to two new factors, both vital for the research-designer to appreciate.

It is odd but true that most of us have an irrational mental block on the subject of high numbers; we tend to call them infinite and then run away from them. There are perhaps two reasons for this. The first is that we find it impossible to count up to high numbers; by the time we have counted, say, 3,000 things we have exhausted ourselves and spent a lot of time. If we are limited to counting, only low numbers become practicable and a million is, in effect, equivalent to the infinite. Both are unascertainable, in any practical sense. But this tendency to banish large numerical values to the limbo of the infinite is based not only on the primitive difficulty of counting but also on the more cultured atmosphere of the arts; for we are all inveterate poets in our attitude to this.

'Coo! Are you having a side-swipe at other cultures?'

I didn't know you were still there; of course I am not having a go at poets; poetry has, is, and will be responsible for more human happiness than science ever can. In practice, if science makes something ten times better it generally makes something else ten times worse, but I am far too discrete to say so. As you have doubtless perceived I am the soul of tact!

'I'd rather not comment on that, but do go on about poetry.'

I'd better define it first and I will quote[20] Professor C. S. Lewis on the subject. He defines it ' As writing which arouses and in part satisfies the imagination'. He goes on to say 'I take it there are two things the imagination loves to do. It loves to liberate its object completely, to take it in at a single glance and see it as something harmonious, symmetrical and self-explanatory. That is the classical imagination: the Parthenon was built for it. It also loves to lose itself in a labyrinth, to surrender to the inextricable. That is the romantic imagination: the Orlando Furioso was written for it.'

I believe we all, researcher included, have an imagination that 'loves to lose itself in a labyrinth, to surrender to the inextricable'. As a result we have an unconscious tendency to dismiss a line of research because we romantically think, and enjoy doing so, that it involves the infinitely big or the infinitely small.

For instance, a celebrated scientist recently stated that he believed that the number of pores (through which leaves breathe) of all the leaves of all the trees of all the world certainly to be infinite.[21] But he was wrong; the number of electrons in a single leaf is much bigger than the number of pores of all the leaves of all the trees in all the world. And still the number of all the electrons in the entire universe can be found by the physics of Einstein and the mathematics of Eddington who put the figure at about 10^{79}.

Now we instinctively feel that there is something unreal or ridiculous about such large numbers but our feelings are not reliable mathematicians.

Actually there are mathematical terms to make big numbers easier to talk about. Ten to the power of a hundred—i.e. one with a hundred zeros after it, is called a 'Google'. An even bigger number, ten to the power of a Google, is called a Googleplex. And to write down a googleplex you would need a piece of paper that went to the farthest star, touring all the nebulae on the way and even then it would not be long enough if you wrote down zeros every inch of the way.

But even a googleplex is too small to measure some things.

For instance, if we imagine the entire universe as a chessboard, and the electrons in it as chessmen, and if we agree that any interchange in position of two electrons is a 'move', then the total number of possible moves would be a googleplex, raised to the power of twenty-four!

Now the figure 1, a google, a googleplex and even this last number are all equally and infinitely far from being infinite.

With the advance of computers I hope we are training up a generation of designers who are not over-awed by the apparently 'infinite' and will only be poetical out of working hours.

The second principle to which the digital registering instrument introduces us is the possibility of mechanizing the abstract. You will remember that for assessing the yield point it was a difficult instrument compared with the others. They indicated what was happening by distances, not digits.

For assessing the modulus of elasticity we found that the digital form was best, since it took us into arithmetic in one step. Thus the 'distance' measuring devices took us one step into the abstract while the digital took us two, in one instrument. We have, in fact, mechanized a part of the abstract process and this opens up fresh possibilities.

10

THE ART OF CONVERSATION

In the last chapter we saw how instrumentation can not only provide a blast off from the concrete into the abstract but provide a second stage as well. This leads us to consider if it would be possible to mechanize, by some comprehensive instrument, all the stages of the abstract and so achieve a splash down on the design level of reality.

Assuming this was possible (and we do not yet know if it is), our research would be greatly simplified in one way but made more difficult in another. It would be simplified in that all we need do is to alter one thing at a time and our hypothetical instrument will then do all the work for us and register the total effect of the modification. The difficulty would be that we would have little or no idea of why it happened. By mechanizing the abstract we would have concealed it. The device would answer the questions beginning with 'what' but never those beginning with 'why'. It would tell us if a thing worked but not how it worked. But, provided we only asked one thing at a time we would not need to know any abstract explanation. We would live on the concrete and keep our heads out of the clouds.

But can such total instrumentation exist? Yes, and in two forms. The first and simplest way of making an instrument duplicate a total machine is to make the total machine itself the instrument. Put in a variable and it hands out a result. You don't have to think. In fact all research for design can be achieved in this way provided that you live long enough and the money doesn't run out.

In the early days of engineering most research was done in this way, and in our childhood days we may well have done something like it too. Are you still with me?

'Yes, but I would like an illustration.'

All right, I will tell you about the first piece of serious research that I ever did.

At school I was compelled to play a very English game called cricket, and I decided to do research into how the game might be livened up a bit. So I invented a new bat. I obtained an aged one and shortened it by cutting about six inches off the bottom and screwed a long strip of sheet lead down the back. The abstract relationship between the weight and position of the lead on the bat and its enlivening effect on the game was quite beyond my abstract reasoning, but this did not matter. I just had to try things out and see. I gradually increased the weight until I obtained the optimum result; all the tests being carried out in school matches. One had, of course, to develop a new technique for hitting the ball. As the new bat was quite heavy it could not be steered side-ways very readily when in full flight, and so one came up the pitch to hit everything full toss. If you missed, the bat, with quite unstoppable momentum, would describe an arc at arm's length and there was a danger that it might descend vertically on top of the wicket. When it did so the bails were smashed and one or more of the stumps driven an unbelievable distance into the ground. The high point of my scholastic career was to hit the first four balls of a match out of the ground, each hit having added psychological effect as the lead, indifferently screwed on, tended to make a peculiar noise of its own. The whole performance tended to produce in earnest cricket enthusiasts a high degree of rage and frustration.

'Another illustration of your being the soul of tact, I suppose?'

I am not commending my behaviour but reporting it as an example of *ad hoc* research by varying one thing at a time until you find the optimum; no thought, all action. Edison tried out in turn 1,600 different filaments before he found the one that worked in an electric light.

Earlier we were talking about axle-tramp and the causes of it. But you need not necessarily worry yourself about the cause at all. Try locating the axle by different methods and test them on the road in turn. You don't have to do mathematics, just listen.

You can picture this process, in general, as a conversation. You say to your instrument-machine 'Is this any good?' 'No' it

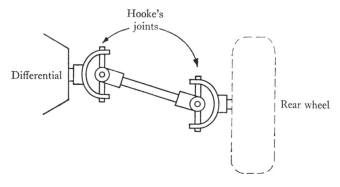

Fig. 11. Diagrammatic drawing of rear transmission with two Hooke's joints.

says. 'This one any better?' 'No, it's worse.' 'How about this?' 'That's more like it', and so on until the pair of you have found the best solution. But, as we have already suggested, such limited thinking can only be bought with unlimited time.

Imagine, for instance, that your particular solution for axletramp has been that of adopting an independent rear springing system, which involves two half axles with Hooke's joints at the ends of each, as shown in Fig. 11.

To conjure up a picture of how this is going to work as a transmitter of the power of the engine, we must first consider how a car will behave when driven by a shaft through a single Hooke's joint.

A Hooke's joint is not a constant velocity one and so, if the driving shaft turns at a constant angular velocity, that of the wheel will vary. But if the shaft is driven from a power source matters become more complicated. If the power generation involves a good deal of inertia, the wheel being light has much less, and the latter will have to absorb the majority of the angular variations. If, however, the wheel is heavy, or has the accumulated inertia of the car reacting through the road surface helping it, the variation in angular velocity will be forced back into the power source; all this is influenced by the angular deflection of the joint at the particular moment. Now if we add on the second

85

Hooke's joint and have three shafts, none in line with any other and each contributing its inertial effect too, we are in for some involved gymnastics.

Alter one thing at a time, measure its effect on the total set-up, and you will be able to find the overall relationships in about ten years provided you work weekends. It will be a long conversation.

This brings us to the second form in which a total instrumentation can be evolved. Perhaps we had better introduce the idea by defining, in general terms, the two essential elements that all such instrument-machines must possess.

The first is that there must be a one-to-one correspondence between each component, force, or velocity in the real machine and that in its instrumental equivalent.

The second is that it must share the characteristic that we demand from all instruments: precision, objectivity, and no need for guesswork.

These characteristics are automatically there when we use the machine itself as its own instrument. But can we duplicate them in some other form.

'How about a computer?'

Yes, you are perfectly right, but we must be very careful that it conforms to our conditions. For instance, you could not alter one thing at a time unless there was an exact and corresponding representation of it in the computer. Take for instance, our double Hooke's joints. The measurements, velocities and so on can be precisely isolated and defined, and their relationships established without guessing. Program a computer to suit and you can have a high speed cross-talk act in confidence, for you have 'ghosted' the actual mechanism.

In the basic picture of our concrete–abstract–concrete loop we can define our instrument-machines or their computer counterparts sitting directly on top of the design area thus giving a one-to-one correspondence and only one abstract step between them.

With this I think I have exhausted my general principles and examples for research design.

'That's a relief anyway, you have certainly exhausted me. Do I have to carry it all in my head and churn it over before I have researched into anything?'

It's not quite as bad as that, as I can compress it into a series of conclusions and recommendations.

11

CONCLUSIONS AND
RECOMMENDATIONS

For reasons that I have outlined in another book, I always advise designers to state their facts and ideas in drawings and diagrams if they can, for words and figures often involve another abstract step. Possibly I may be wise to accept my own advice and so I have prepared a single diagram (Fig. 12) incorporating my conclusions, and my recommendation is that you should have a good look at it.

'It looks to me like a map of the London Underground gone mad.'

Well, in a sense, it is a map, for it outlines the basic mental and research routes that a designer can follow when searching for data. In a sense it is an index too, for the numbers in circles are those of the chapter in the book dealing with that particular part of the map.

'What use do you think all that will be?'

I suggest that the first thing you do, when given the responsibility of finding out by research some missing facts that you need for designing, is to take a long hard look at this diagram. It contains, in principle, all the options open to you.

'What can I do if I don't know enough about a problem to decide in advance where to put it on the map?'

You must do some preliminary research into what is going on. It is surprising how, sometimes at least, you can play about with bits and pieces of things and so quite quickly get the general feel of what it is all about. Don't rush out and buy expensive instruments before you are satisfied that they are essential. Don't be bull-dozed by yourself, or anyone else, into a research programme until you know where you are, and where you are going. And the third don't is: don't lose your nerve. Having decided on a programme, act on it. Standing there and looking at it will ruin your confidence and madden your employer. You may well be better at doing research than you think, so have a

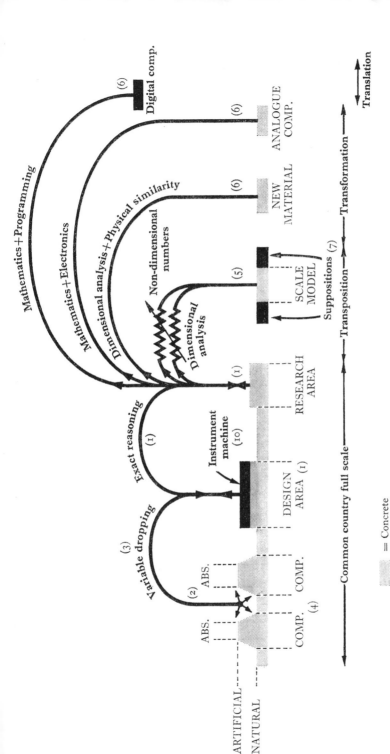

Fig. 12

look at that diagram, make up your mind and plunge into the adventure of doing it. There are few things more rewarding than contemplating a finished and working engineering design that you have created through disciplined thinking and logical research. If it actively benefits the community as well, you will never lose this sense of satisfaction.

REFERENCES AND NOTES

1 *Making and Interpreting Mechanical Drawings.* G. L. Glegg. C.U.P., 1971.
2 *Designing against Fatigue*, page 13. R. B. Heywood. Chapman & Hall, 1962.
    You will also find the following useful:
    Residual Stresses and Fatigue in Metals. John O. Almen & Paul N. Black. McGraw Hill, 1963.
    Thermal and High Strain Fatigue. The Metals and Metallurgy Trust, 1967.
    Fatigue Testing and Analysis of Results. W. Weibull. Pergamon, 1960.
    The Brittle Fracture of Steel. W. D. Biggs. Macdonald & Evans, 1960.
    Fatigue of Welded Structures. T. R. Gurney. C.U.P., 1968.
3 The following books are helpful on Creep:
    Creep and Stress Relaxation in Metals. I. A. Oding (ed.). Oliver & Boyd, 1965.
    Plasticity for Mechanical Engineers. W. Johnson & P. B. Mellor. Van Nostrand, 1962.
    Creep of Concrete, Plain, Reinforced and Prestressed. A. M. Neville. Elsevier, 1970.
4 Response of Committee of Enquiry into Collapse of Cooling Towers at Ferry Bridge 1965. C.E.G.B. (1966).
5 *The Design of Design.* G. L. Glegg. C.U.P., 1969.
6 Quoted from: *Machine Tool Dynamics – An Introduction*, page 12. D. B. Welbourn & J. D. Smith. C.U.P., 1970.
7 *Principles of Structural Analysis.* T. M. Charlton. Longmans, 1969.
8 See fuller description in: Aerodynamic Testing of Vehicles at Mira. R. G. S. White & G. W. Carr. *Proc. I. Mech. Eng.* vol. 182, part 3B, page 65.
9 *The Selection of Design.* G. L. Glegg. C.U.P., 1972.
10 *Use and Abuse of Statistics.* W. J. Reichmann. Methuen, 1961.
    Statistical Methods for Technologists. C. G. Paradine & B. H. Rivett. English Universities Press, 1970.
    An Introduction to Statistical Methods. H. J. Halstead. Macmillan, 1961.
11 *Engineering Applications of Digital Computers.* T. R. Bashkow. Academic Press, 1968.
12 These books are amongst many that are helpful:
    Engineering Plasticity. G. R. Calladine. Pergamon, 1969.
    Dynamics of Mechanical Systems. J. M. Prentis. Longmans, 1970.
    Mathematics of Engineering Systems. D. F. Lawden. Methuen, 1959.
    Essentials of Control Theory for Mechanical Engineers. D. B. Welbourn. Arnold, 1963.
    Mechanical Vibrations. D. den M. Hartog. McGraw Hill, 1956.
13 These are all useful, but there are many others:
    Beams and Framed Structures. J. Heyman. Pergamon, 1964.
    Matrix Methods of Structural Analysis. R. K. Livesley. Pergamon, 1964.
    Mechanics of Materials. F. R. Shanley. McGraw Hill, 1967.
    Plastic Design of Frames. J. F. Baker & J. Heyman. C.U.P., 1969.

The Steel Skeleton Vol. II. J. F. Baker, M. R. Horne & J. Heyman. C.U.P., 1956.

Braced Frameworks. E. W. Parkes. Pergamon, 1965.

14 *Dimensional Analysis and Scale Factors.* R. C. Pankhurst. Chapman & Hall, 1964.

Modern Developments in Fluid Dynamics. S. Goldstein (ed.). Dover, 1938.

An Introduction to Fluid Dynamics. G. K. Batchelor. C.U.P., 1967.

Experimental Fluid Mechanics. P. Bradshaw. Pergamon, 1964.

Wind Tunnel Technique. R. C. Pankhurst & D. W. Holder. Pitman, 1968.

Hydrodynamics in Theory and Application. J. M. Robertson. Prentice-Hall, 1965.

15 *Models for Structural Concrete.* B. W. Preece. CR Books, 1964.

Model Analysis of Plane Structures. T. M. Charlton. Pergamon, 1966.

16 *Analogue Computation in Engineering Design.* A. E. Rogers & T. W. Connolly. McGraw Hill, 1960.

17 *An Introduction to Automatic Digital Computers.* R. K. Livesley. C.U.P., 1960.

18 All this is explained better and more fully in: Performance characteristics of Roots blower systems. B. N. Cole & B. W. Imrie. *Proc. I. Mech. Eng.* vol. 189, part 3R, page 116.

19 Failure of a 60-MW steam turbo-generator at Uskmouth power station. A. G. Lindley & F. H. S. Brown. *Proc. of I. Mech. Eng.* vol. 172, page 627.

20 *They Asked For a Paper*, page 151. C. S. Lewis. Bles, 1962.

21 I have borrowed freely from that fascinating book, *Mathematics and the Imagination*. E. Kasner & J. Newman. Pelican, 1968.

INDEX